Recruiting, Training, and Maintaining

Volunteer Firefighters

THIRD EDITION

Jack W. Snook
President
Emergency Services Consulting Inc.
Wilsonville, OR

Jeffrey D. Johnson
Fire Chief
Tualatin Valley Fire and Rescue
Aloha, OR

Dan C. Olsen
Fire Chief
South Lane County Fire & Rescue
Cottage Grove, OR

John M. Buckman III
Special Projects Manager
Division of Training
Indiana Department of Homeland Security
Indianapolis, IN

JONES AND BARTLETT PUBLISHERS
Sudbury, Massachusetts
BOSTON   TORONTO   LONDON   SINGAPORE

**International Association
of Fire Chiefs**

Jones and Bartlett Publishers
World Headquarters
40 Tall Pine Drive
Sudbury, MA 01776
978-443-5000
www.jbpub.com

Jones and Bartlett Publishers Canada
6339 Ormindale Way
Mississauga, ON L5C 2W6
Canada

Jones and Bartlett Publishers International
Barb House, Barb Mews
London W6 7PA
United Kingdom

**International Association
of Fire Chiefs**
4025 Fair Ridge Drive
Fairfax, VA 22033
www.IAFC.org

Jones and Bartlett's books and products are available through most bookstores and online booksellers. To contact Jones and Bartlett Publishers directly, call 800-832-0034, fax 978-443-8000, or visit our website www.jbpub.com.

Substantial discounts on bulk quantities of Jones and Bartlett's publications are available to corporations, professional associations, and other qualified organizations. For details and specific discount information, contact the special sales department at Jones and Bartlett via the above contact information or send an email to specialsales@jbpub.com.

Production Credits
Chief Executive Officer: Clayton E. Jones
Chief Operating Officer: Donald W. Jones, Jr.
*President, Jones and Barlett Higher Education
and Professional Publishing:* Robert W. Holland, Jr.
V.P. of Sales and Marketing: William J. Kane
V.P. of Production and Design: Anne Spencer

V.P. of Manufacturing and Inventory Control: Therese Connell
Publisher, Public Safety Group: Kimberly Brophy
Acquisition Editor: William Larkin
Production Editor: Jenny McIsaac
Director of Marketing: Alisha Weisman
Text Printing and Binding: Odyssey Press, Inc.
Cover Printing: Odyssey Press, Inc.

ISBN: 0-7637-4210-4

Library of Congress Cataloging-in-Publication Data not available at time of printing
6048

Printed in the United States of America
10 09 08 07 06   10 9 8 7 6 5 4 3 2 1

This Book

is

Dedicated

to the Memory of

Wes Claflin

Volunteer Fire Captain

Jackson County (Oregon) Fire District No. 3

1961 - 1979

and

All Volunteer Fire and Rescue Personnel

throughout America who give

unselfishly of themselves.

Captain Claflin was what being a Volunteer Firefighter is all about:

"A teacher, a friend, a person who cared about people."

Acknowledgments

Clackamas County Fire District No. 1, Oregon
Oregon Volunteer Firefighters Association
Washington State Fire Service Training
Rogue Fire Training Association, Oregon
California Fire Service Training and Education
South Willamette Training Association, Oregon
Pierce County Washington Chiefs' Association
Pierce County Fire District No. 5, Washington
Tahoe City Fire Department, Washington
Cottage Grove Fire Department, Oregon
Klamath Falls Fire Department, Oregon
Lacey Fire Department, Washington
Happy Valley Fire Protection District, Oregon
Jackson County Fire District No. 3, Oregon
Lake Oswego Fire Department, Oregon
Beaverton Oregon Fire Department, Oregon
Tualatin Valley Fire and Rescue, Oregon
Fairfax County Fire and Rescue, Virginia
German Township Indiana Volunteer Fire Department
International Association of Fire Chiefs
Volunteer Section of the IAFC
National Volunteer Fire Council
Volunteer Fireman's Insurance Services
Oregon Volunteer Fireman's Association
Wayne Township Indiana Fire Department
Loudon County Virginia Fire and Rescue
Prince George's County Maryland Fire Department
Farmington Hills Michigan Fire Department
Ponderosa Texas Volunteer Fire Department
Boone County Missouri Fire Department
Volusia County Florida Department of Fire Services
Montgomery County Virginia Department of Fire and Rescue
New Jersey Volunteer Firefighters Association
Knight Township Indiana Volunteer Fire Department
Sara Smith
Carlton Williams

Technical Assistance Provided By:

Barry Enoch, Robert Weiderhold, Randy Bruegman, Garry Briese

Illustrations By:

John Schoffner
Terry Stoltz

Also, a special thanks to
Sharon Sayler
at Wild Rose Design

Table of Contents

Introduction

Volunteer Firefighters

Volunteer firefighters have been the backbone of the fire service for over three hundred-sixty years. They have been a part of much change during this time period. They have trained only to be retrained. They have learned only to relearn. No volunteer group, in the history of this country, has had to work so hard and sacrifice so much as have the volunteer firefighters.

Volunteers will be experiencing additional difficulties in the coming years. Increased requirements for time, family, work, and the economic climate will challenge the leader who uses volunteers in their delivery system.

As we face increasingly difficult economic times the need for volunteers and "volunteerism" magnifies. Organizations which have historically counted on volunteers are working diligently to maintain their delivery systems. Departments which have relied primarily on paid firefighters are looking towards volunteers to support their paid forces and fill in the "gaps" being created by financial shortfalls. Hopefully, once we analyze the service that we provide, the needs of the community as well as the needs of those individuals who volunteer their services, we will find many answers to questions that have become of critical importance during these past few years. To identify alternatives and develop solutions to the problems, we must first understand the nature of the situation, and the complexity of the environment as it relates to utilizing volunteer firefighters. This was not written to assist you in merely surviving, it was written to enhance fire protection and related services. In reviewing this material, one must realize that it is just that: it is written material. Not until we commit ourselves, and put ideas, concepts, and techniques to work, shall we see positive results within our organizations. This takes a commitment from you and everyone within your department.

A successful volunteer fire department will have highly effective leadership. The single most important ingredient to success is "dynamic and effective leadership." "Success has a high price to pay. The successful volunteer program will have paid the price."

The volunteer fire department will face many challenges in the 21st century. Commitment and determination will be necessary to maintain and improve the delivery of service to the customer.

The volunteer fire department of tomorrow will not only be in the fire and E.M.S. business, but the customer service business as well. Customer service is determined by what the customer wants at a specific time. The fire department is the one government service that delivers to the public a wide variety of services at any time of the day or night. Fire in many cases will seem to become a sideline because of the other services that the public will demand. Services such as emergency medical services, vehicle rescue, confined space rescue, high angle rescue, hazardous materials, or even assistance with water, sewer, and power outages may be the priority at any given time.

No volunteer group, in the history of this country, has had to work so hard and sacrifice so much as have the volunteer firefighters.

Success has a high price to pay. The successful volunteer program will have paid the price.

One of the highest motivators for volunteer firefighters is emergency response. The challenge of the leader is to educate and motivate the volunteer that the primary mission is to help people. Whatever type of service the customer wants is what the fire department will have to be ready to deliver.

Facts About Volunteerism In America

"This form of public service embodies the American values of democracy, patriotism, and individual freedom." (Jacobs, 1976)

- Three-fourths of the United States is protected by volunteer firefighters
- There are approximately 1,000,000 volunteer firefighters in the United States
- Volunteers donate approximately 65 million hours each year in the United States
- Volunteer time is valued at $105 billion
- There are approximately 27,000 volunteer fire departments in the United States
- There is a 3% annual increase in volunteers in the United States
- Volunteerism is on the upswing; it is a FALLACY that volunteerism is dying
- Over 200 new volunteer organizations are formed each day in the U.S.

Famous Volunteer Firefighters In America

- George Washington – Thomas Jefferson
- Ben Franklin – Samuel Adams
- John Hancock – Paul Revere
- Alexander Hamilton – John Ray
- Aaron Burr – Benedict Arnold

Distinguishing Features of Volunteer Firefighters

- Traditional by Nature
- Grassroots Origin
- Autonomy – "don't tread on me"
- Service Monopoly
- Solidarity – "trust and team work"
- Role Performers
- Focus for Leisure (Jacobs & Lozier, 1976; Perkins, 1987a, 1987b)

Definition Of Terms

Voluntary — To act out of one's own will and choice.

Voluntarism — A generic term for that which is done voluntarily.

Volunteerism — The umbrella term for all that is done by volunteers.

Voluntary Associations — Groups which are developed by and comprised of volunteers, sometimes with the help of paid staff, but for no profit.

Professional — Having special training and skills; conducting oneself in a manner respected by members of the profession and society in general; engaged in something worthy of high standards not necessarily for monetary gain.

"This form of public service embodies the American values of democracy, patriotism, and individual freedom."

Jacobs, 1976

History and Background

Volunteerism

Volunteerism in America had its beginning as early as 1620. It was at this time that the New World established a cooperative arrangement which achieved predetermined goals and objectives. This first attempt at a voluntary effort was accomplished through the "Social Compact" of 1620. In this document the Pilgrims affirmed the necessity for a government based upon the consent of the governed. They further joined in a covenant which bound them "strictly tied to all care of each others good and of the whole by everyone and so mutually."[1] Early volunteer efforts included: farmers helping each other with harvest; town meetings to discuss important community issues; church organizations which not only constructed their own buildings but assisted members whenever there was a need; and volunteer committees which helped during time of epidemics. Even education was handled on a volunteer basis prior to support from local government and states.

In 1702 the first social and fraternal orders were developed. One of the first recorded was a Masonic Lodge established in Philadelphia in 1715. As time went on, however, many clubs formed along craft and professional lines for the purpose of sharing knowledge and information were cast aside for fellowship in local taverns and private homes.

Later it became apparent that there was a genuine need for public facilities, programs, and services. The need for travel encouraged the development of roads. Thus, volunteer groups were organized to assist in their construction and maintenance.

A first step towards a national postal service was initiated in 1639. This effort in Massachusetts was originally handled by volunteers. In 1677 the Mayor of New York commissioned six wells dug. As there were no public funds available the project was completed by volunteer efforts. The growth of towns brought with them the need for emergency response groups. One of the first was the volunteer fire department. Benjamin Franklin started the first American fire brigade in Philadelphia (1736). By the late 1700's many volunteer groups had been formed in order to meet the needs of growing communities. As people began looking west there was even the establishment of a group of volunteers who would brave the frontier and lead settlers west. Once they arrived, all parties were obliged to assume volunteer responsibilities such as judge, militia commander, surveyor, doctor, teacher, and any other community service which they had a knack to handle or provide.[2]

Beginning in the mid-1700's politics became a daily concern for colonists. As the crisis amongst the colonies grew, so grew the need to link communities, townships, and assemblies. An outgrowth of this situation was the "Committees of Correspondence." Although there was danger in being a member, many citizens participated in this early political volunteer effort in order to share vital information as to local conditions.

The boundaries between public and private interests blurred for a time as the new communities took root. Settlers "...obligate(d) themselves to aid each other, so as to

Volunteerism in America had its beginning as early as 1620. It was at this time that the New World established a cooperative arrangement which achieved predetermined goals and objectives.

The growth of towns brought with them the need for emergency response groups. One of the first was the volunteer fire department. Benjamin Franklin started the first American fire brigade in Philadelphia (1736).

The Civil War brought with it the need for both sides to rally around their soldiers. The need and importance of volunteers had never been more apparent or necessary.

The great depression of the 1930's challenged the American people like no other time period. With widespread unemployment, hunger, and homelessness, citizens rallied behind one another to elevate volunteerism to another level in American history.

make the individual interest of each member the common concern of the whole company."[3] Following the early colonial pattern, neighborliness was an expected part of frontier life. For example, the arrival of a new family brought the settlement together for a cabin raising. Describing the Ohio and Mississippi Valley settlements of 1815, Timothy Smith remarked: "The people more naturally unite themselves into corporate unions, and concentrate their strength for public works and purposes." Timothy Flint also observed...a more familiar, and seemingly a more cheerful intercourse between the two sexes than in other western states."[4]

To protect New York from disease there was established the first health committee in 1794. During this period there emerged a new concept, public health defined as: "... that responsibility which rests upon the community for the protection of life and the promotion of the health of its people. It is one of the basic community functions...recognition of the necessity for community action in promotion of individual and community health and welfare."[5]

The Civil War brought with it the need for both sides to rally around their soldiers. The need and importance of volunteers had never been more apparent or necessary. This volunteer effort would prove to be critical, not only in the reconstruction of the nation, but to the future approach our country would take towards the need for citizens to step forward and assist within their communities.

The last quarter of the nineteenth century saw the development of many secret fraternities. Although initially formed for "men only," fraternities for women also emerged. Issues being addressed by these secret fraternities ranged from farm issues to women's rights.

At the same time there continued to be major emphasis put on volunteer efforts in medicine. The best known volunteer organization established after the Civil War was the American Red Cross. By 1900 it had been granted a Congressional charter. From the beginning, Clara Barton felt that if the Red Cross societies were to be of use in time of war, they should train volunteers to handle natural disasters in time of peace. Thus, their first tests came in the form of devastating forest fires in Michigan (1881) and floods along the Ohio and Mississippi Rivers (1882, 1884). Red Cross relief was given to victims in these and other cases. "The idea of an organized program of voluntary relief for disaster victims was the unique contribution of Clara Barton and the Red Cross movement world wide."[6]

In the wake of World War I, America turned towards growth and the fulfillment of "The American Dream." The pursuit of wealth was a major preoccupation of most Americans. President Hoover was optimistic about the citizens' ability to rally against severe economic hardship. He later described the nature of these efforts: These committees had no politics. They were men and women experienced in large affairs, sympathetic, understanding of the needs of their neighbors in distress. And they served without pay. In those days one did not enter into relief of his countrymen through the portals of a payroll. American men and women of such stature cannot be had as a paid bureaucracy, yet they will serve voluntarily all hours of the day and defer their own affairs to night.[7]

The great depression of the 1930's challenged the American people like no other time period. With widespread unemployment, hunger, and homelessness, citizens rallied behind one another to elevate volunteerism to another level in American history. Soup kitchens and bread lines were found in almost every community.

Since 1945 volunteer efforts have continued to expand. We still have traditional organizations such as the American Red Cross, volunteer fire departments, church groups, etc., but we also have seen the emergence of literally tens of thousands of groups intended to assist, affect, or enhance the livability of our homes, neighborhoods, communities, counties, states, and even the nation itself. In order to truly appreciate the importance of the volunteer effort today, and in the future, we must first gain an appreciation and respect for the efforts made in the past.

[1] Susan J. Ellis, **By the People**, (Philadelphia: Prestegord & Co., 1978), 8.

[2] Susan J. Ellis, **By the People**, (Philadelphia: Prestegord & Co., 1978), 28-31.

[3] Wilson G. Smillie, **Public Health, Its Promise for the Future**, (New York: The Macmillan Co., 1955), 6.

[4] Roscoe L. West, **Elementary Education in New Jersey: A History**, (New Jersey: D. Van Nostrand Co. Inc., 1964), 10.

[5] Smillie, op. cit., p.1.

[6] John Allen Krout and Dixon Ryan Fox, **The completion of Independence** (New York: The Macmillan Co., 1943), 322.

[7] **Civil Air Patrol News, VIII**, No. 2. March, 1976. p.3.

Since 1945 volunteer efforts have continued to expand.

Notes ■

Why Volunteer Firefighters?

Economics

Prior to actively recruiting volunteers, a person must address several questions. First, we must ask ourselves: why volunteer firefighters? The primary reason has been, and will continue to be, economics. One can pick up any newspaper and with little effort can scan various articles and see that public funds are quickly becoming a commodity of great demand and short supply.

Cities and fire districts alike are looking at the utilization of volunteers to assist in maintaining or increasing levels of service while financial resources are being reduced. There is little doubt in the minds of most fire administrators as to value of highly trained paid crews; unfortunately, with the economic conditions we are now experiencing, it appears impossible for medium to small districts and cities to hire sufficient numbers of paid firefighters to meet all community needs, both emergency and non-emergency. More and more volunteers are being called upon to provide that portion of the delivery system which has traditionally been the genius of paid firefighters. That is to respond quickly, assess a situation, implement a plan of action, provide supervision and leadership, and perform tasks. Volunteers are being taxed as are paid firefighters. They are required to do more and more while, in many cases, given less and less to do it with. It was always considered asking a lot of volunteers to have them support paid crews and assist in handling major alarms. Now they must do much more. In some cases, they are the only component present, thus emphasizing, even more, the need to be proficient, dedicated, confident, and competent.

Cities and fire districts alike are looking at the utilization of volunteers to assist in maintaining or increasing levels of service while financial resources are being reduced.

Other Reasons For Having Volunteers Include:

- Volunteers enable organizations to expand or improve services.
- Improves public relations.
- They share the experiences and wisdom from their private lives with the fire department.
- Increases labor pool for special activities.
- They provide for a better community understanding of financial needs.
- It is the only available alternative in many communities.

Butterflies May Be Free – But Volunteers Are NOT!

Webster's dictionary defines "free" as "without charge." Volunteers are not free.

Let's take a look at the volunteers who are the cornerstone of the American Fire Service. They are highly trained professionals who donate their time and expertise to protect and serve our neighbors in the event of a fire, medical or other emergency.

They must complete exactly the same training and certification as paid firefighter counterparts and, in fact, many volunteers are "career" (paid) firefighters with other

Webster's dictionary defines "free" as "without charge." Volunteers are not free.

area departments. Volunteer firefighters must also meet exactly the same continuing education standards as their paid counterparts. They need to act just as quick and efficiently during an emergency. Everything they do is the result of well-learned procedures and techniques that enable them to put the fire out and to stay safe while they are doing it.

There is no such thing as a "volunteer fire" or a "paid fire." No citizen in need of professional emergency assistance asks whether the responder is "volunteer" or "paid." Citizens do not ask to see a firefighter's credentials before allowing them on their property. An emergency is an emergency no matter who responds...something firefighters must be ready to do 24 hours a day, 365 days a year. Your labor force may volunteer their time, but it costs the same to outfit volunteer firefighters as it does paid personnel. Like any other small business, volunteer fire departments must also budget for facilities, equipment purchase and maintenance, insurance, utilities, fuel, training, continuing education, office supplies and so on. Fortunately for most small businesses, however, they don't have to worry about capital equipment replacement cost of $350,000 for a Class A engine or $750,000 for a ladder truck.

Volunteers may provide services without an hourly "salary," but they are paid in other currency. Most paid personnel who have never been a volunteer do not understand the motivation of volunteers. Volunteers require substantial compensation for the job they do; this "payment" is provided in the form of both tangible and non-tangible awards and incentives.

The safety equipment worn by volunteers is to provide for increased personal protection when facing the dangers of fire and related emergencies. This "investment" comes at a significant cost. Outfitting just one firefighter in basic, required gear costs a staggering $2,000 or more. Add to that the cost of the air pack that each firefighter must wear and the cost can exceed $5,000.

Volunteers spend many hours preparing for emergencies. Continuing education and training is provided in fire suppression, emergency medical services, hazardous materials, urban rescue, water rescue and automobile rescue to name just a few.

It is an unfortunate perception that "volunteer" is synonymous with "free," a thought that could jeopardize our future ability to provide the extraordinary level of services communities expect and desire.

Operating costs in many communities outstrip funding resources and compromises will unfortunately become necessary. The compromises will NOT be with the health and safety of volunteers, but may well impact the number of volunteers they can support and the equipment they use, or the number of fire stations we can afford to operate.

The cost of replacing a current volunteer force with paid personnel is something that many communities cannot realistically afford to do. In order to ensure the

There is no such thing as a "volunteer fire" or a "paid fire." No citizen in need of professional emergency assistance asks whether the firefighter is "volunteer" or "paid."

Volunteers may provide services without an hourly "salary," but they are paid in other currency.

continued viability of the volunteer fire service it is important to explore providing adequate benefits to the professional volunteers. Benefits can range from free uniforms, free advanced training, stipend reimbursement for responding to emergency calls, training or meetings, length of service award programs, pensions, personal property tax breaks, etc. The type of benefit is limited only by the creativity of the community that desires to maintain the volunteer system.

Cultural differences in a variety of communities across America make it difficult to ascertain the exact actual cost of a volunteer firefighter. In many communities, volunteer firefighters may actually be paid, on-call personnel who receive cash compensation while responding to emergency calls. Compensation of volunteers requires considerable study prior to implementing. Once implemented it will be difficult if not impossible to stop. There are a variety of forms of cash compensation: pay per hour, pay per call, pay based upon a lump sum of budgeted moneys. Members can be paid for a variety of activities other than responding to emergency calls, such as training, standby time, maintenance and public events.

There is no reason professional volunteer organizations cannot survive and thrive in America today. The resources necessary for this to happen will have to be explored and expanded.

The next time someone says "let a volunteer do it," remember that while a professional volunteer may be less expensive, they are certainly not free.

What Is The Job Of A Volunteer Firefighter/Department?

Volunteer firefighters are asked to perform the same tasks as their career counterparts. Fire departments provide a wide variety of services that the citizens of the community request. Listed below are some of those types of services.

- Fire suppression
- Emergency medical services

    First responder – non-transport

    Basic life support – with transport capability

    Advance life support – with transport capability

- Hazardous materials response

    Operations level

    Technician level

- Specialized rescue

    Auto extrication

    Confined space rescue

    Trench rescue

    Water rescue

    High angle rescue

- Public information and education

    Fire prevention programs for adults and children

    Public presentations for a variety of community organizations

Compensation of volunteers requires considerable study prior to implementing. Once implemented it will be difficult if not impossible to stop.

The next time someone says "let a volunteer do it," remember that while a professional volunteer may be less expensive, they are certainly not free.

- Inspection services
- Fund raising activities

    Annual events (dinner, firefighter's ball)

    Special events (raffles, garage sales)

    Weekly activities (bingo)

    Fund drives

Each one of these services has an impact on the volunteer firefighter. The delivery of each of these services carries along with it another commitment from the community that includes paying for the cost of those services. Cost of providing them can be an economic burden on the taxpayers. Each fire department and community must realistically examine their needs and plan accordingly.

The expensive part of providing each of these services is not necessarily in the delivery but is in the preparation and planning. Preparation and planning includes the purchase of the equipment as well as the training commitment required of the volunteer members.

Most fire departments who deliver some level of emergency medical related services will find that the vast majority (60%+) of their emergency calls will be for medical situations. It is difficult to measure the effectiveness of emergency medical services in the number of lives saved but it is not difficult to measure in the number of people helped.

When a customer dials 9-1-1 and the firefighter is the first one to show up at their front door they will always remember who was there to help when they needed it.

Although many fire departments resist becoming involved in EMS it is becoming a necessity with fires being reduced through better codes, technology, better construction, and public awareness.

Many citizens of communities do not understand the high cost of operating a professional volunteer organization. It is expensive to equip and operate a functioning emergency services organization.

Can Volunteers Do The Job?

The second question commonly asked is: can volunteer firefighters provide or assist in providing needed community service? In order to analyze this we must take a look at department goals. Listed below are common goals recognized throughout the fire service.

Fire Department Goals

1. To provide performance levels acceptable to the citizens residing within the jurisdictional boundaries of a fire department.

2. To provide life safety levels acceptable to the citizens residing within the jurisdictional boundaries of a fire department.

3. To confine fires to a modular level acceptable to the citizens residing within the jurisdictional boundaries of a community.

4. To suppress fires with the least amount of property damage.

5. To reduce large fire frequency.

6. To provide selective emergency services desired by the community. (EMS, public education, etc.)

7. To meet performance levels that have been established under a favorable ratio of the cost to performance effectiveness.

8. Reduce the total number of incidents within the community.

Problems Utilizing Volunteers

Can volunteers meet these objectives? The answer to this question is simply and emphatically YES. Do problems arise? The answer to this question is definitely YES.

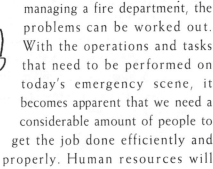

There are going to be problems utilizing volunteer firefighters within a delivery system, however, working together, utilizing a "systems" approach to managing a fire department, the problems can be worked out. With the operations and tasks that need to be performed on today's emergency scene, it becomes apparent that we need a considerable amount of people to get the job done efficiently and properly. Human resources will dictate, to a large degree, success or failure. Volunteers will continue to be a critical resource in providing needed staffing.

Even the best volunteer programs face many challenging situations. Any time volunteers are used there arise potential problems. These problems can be categorized in two different areas: physical and psychological. Physical problems include availability of volunteers, attrition, lack of training time, and lack of blue collar workers to recruit. These problems for the most part are unavoidable and must be coped with for they are difficult to eliminate. Psychological problems, on the other hand, such as attitudes, morale and a feeling of accomplishment through dedication and team work, are areas which most definitely should be addressed and handled effectively in a good volunteer program.

There is a tremendous need for people during the first few minutes of most structural fire problems. Few departments can respond and commit personnel to handle necessary tasks in a systematic, timely fashion. Let's take a look at a common structure fire: A three bedroom residential home which has several rooms involved. To properly handle this situation, we would need to perform several tasks which would include advancing initial attack lines, advancing and staffing back-up lines, operating pumpers, commanding the fire, providing forcible entry, providing for water supply, and taking care of salvage and overhaul once the incident is stabilized. As we look at these operations, we come up with a staffing requirement of somewhere between nine to twelve firefighters to properly handle this rather common fire problem. We must ask ourselves at this point whether we can, on a day-to-day basis, provide this kind of response.

There are going to be problems utilizing volunteer firefighters within a delivery system; however, working together, utilizing a "systems" approach to managing a fire department, the problems can be worked out.

We quickly realize there is a tremendous need for human resources on today's emergency scene.

Once this is done, we must then ask ourselves just how many people can be put on the scene and what the capabilities are to handle larger incidents, as it relates to what's at risk in the community.

Then we quickly realize there is a tremendous need for human resources on today's emergency scene. If we cannot provide such staffing, through the utilization of paid firefighters, volunteers, and/or a combination of the two, where do we stand? Many studies have been done regarding the effectiveness of teams of firefighters to the number of persons within a specific group. For example, let's assume that a six person engine company is 100% effective as it relates to being trained to do specific tasks within a specified time period. As we reduce the number of persons within this team, we find some unusual problems and circumstances arising. For example, reduce to five people and they become approximately 80% as effective as a six person team. Reduce that team to a four person team and they become approximately 50% as effective as a six person team. Notice that we have reduced the staffing by 33% but the effectiveness of the personnel has been cut in half. As we can see, the number of trained personnel that arrive initially, or shortly thereafter, are going to play an important role in the end result which we are trying to achieve. A trained volunteer force can provide the staffing necessary to keep losses at a minimum, effectively handle major emergencies, and provide necessary staffing.

What The Future Has In Store For Volunteer Firefighters

Although no one can look into a crystal ball to determine the future of the fire service, it does appear that more and more emphasis will be placed on "volunteerism." As it becomes more and more difficult to finance various programs within departments, it becomes apparent that volunteers must take on additional responsibility in an effort to maintain service levels. The volunteer firefighter will have to be trained to a higher level, they will need to keep themselves in physical and mental shape, and they will have to give more. This is a tremendous challenge, but one our volunteers will need to meet.

Volunteer firefighters will still be around in the foreseeable future. There is no other group of people who give so much to their communities for so little in return. The volunteer firefighter is an excellent investment for most communities.

Each and every one of us uses the excuse "I am too busy" for a variety of reasons. Many of us tend to not read trade journals, books or any other type of educational related material. "Time" is usually the reason given for not being able to do those things. Everyone in the fire service, especially the volunteer, is under a time crunch. Time management is an important skill that we each need to improve upon. But even with time management skills leaders in the volunteer fire service will have to prioritize their activities if they are going to be effective leaders.

One aspect of fire/rescue department organization that deserves additional attention is the legislative process. The legislative process as I have defined it includes administrative law as well as legislated laws.

Each and every one of us needs to understand that we must pay attention to what is going on in our local legislative body and our state legislative body as well as at the federal level. You cannot leave this important task to someone else.

As it becomes more and more difficult to finance various programs within departments, it becomes apparent that volunteers must take on additional responsibility in an effort to maintain service levels. The volunteer firefighter will have to be trained to a higher level, they will need to keep themselves in physical and mental shape, and they will have to give more. This is a tremendous challenge, but one our volunteers will need to meet.

Not only are the legislative bodies passing legislation that might change the face of your department but there are government bureaucratic agencies that "regulate" that many times are much more painful to our departments than legislation.

To ignore what is going on around us is akin to putting our head in the sand and letting the world pass us by. Although we as chief or a volunteer leader might want to do that, eventually we raise our head up and we find that things have changed and we wonder how this happened. We demand in some cases to know why our state/national organization allowed this to happen. We are furious that the dues paid to the organization did not produce the results that we wanted. The solution... Get involved!

Wonder! Wonder! How Could That Happen?

State legislators don't always listen when an organization attempts to "lobby" their position on particular legislation. Federal legislators don't always care when a national organization representing several thousand members calls the legislator's staff and requests a position on a particular piece of legislation.

It's Always Someone Else's Job!

Today, there are people out there attempting to pass legislation that will dramatically change the face of the fire service. Some of the legislation may actually help us. Some of it may not!

But you know what? If YOU do not pay attention to what your state and federal representatives are doing and become proactive in letting your position be known, then your wishes will not be heard and your desires will not be carried forward.

How can we change the way we look at regulatory affairs? Simply put, it's time to put a position on your organization chart titled "regulatory liaison." This person would be charged with monitoring local, state and federal regulatory legislation that you might want to take a position on.

This position alone will not solve the problems but it might help us address some of those legislative/regulatory issues when we still have a chance of impacting the final product.

We have positions within our organization such as training, fire prevention, apparatus maintenance, etc. Why? Because these are important tasks that your department monitors in order to be effective at delivering service. When we put someone in charge of something, a job description must be developed as well as parameters in which the person may operate. A budget may also have to be developed.

This liaison position would be an important task within your department. There are many organizations that are not governmental agencies that you might also want to be aware of.

Organizations develop standards that your department may be measured against and you may not have known about, and in most cases had very little input into the

Organizations develop standards that your department may be measured against and you may not have known about, and in most cases had very little input into the development of, but they are still out there. Get Involved!

The strongest representative you have is yourself and your members. Only you can speak for you. Ignore it and it will not go away. Take action and you might have an impact.

development of, but they are still out there. OSHA, on both the state and federal level, develops safety standards for the fire service. The National Fire Protection Association (NFPA) establishes all types of standards.

Let me give you some examples of regulatory/legislative actions that have happened and are happening today!

A state fire organization decides to hire a telephone solicitor to solicit throughout the state for funds to finance the organization. Does this action impact you? In most cases, YES. This action will in many cases take money from the local fire department and put it in another organization's hands. Will this solicitation directly help you in providing service to your citizens? How does the state organization know how you feel? The state organization isn't concerned about your fire department's financial needs, they only want or need money to operate. So do you! The result of this action is close to stealing right out of your pocket, and you know what? No one else cares. Most people and organizations look out only for themselves and not for anyone else.

A federal government agency determines that carrying SCBA's in side compartments or in jump seats is unsafe. Regulation is developed and orders that all SCBA be placed in boxes and carried inside compartments. Yes, this is an actual proposal.

Could either one of these scenarios take place? Certainly. Could they be stopped or changed after the decision has been made? Maybe! But unlikely.

The volunteer fire service is great at standing up after the fact and telling people what we don't like about something that has already been passed. We sometimes expend tremendous energy to try changing something after it exists. It is time to become proactive in our response to agencies, government and groups that develop legislation/regulations and let them know during the development process what we think about the proposals. Get involved!

Each fire department must pay attention to local, state and federal regulatory action. It just might be important.

The strongest representative you have is yourself and your members. Only you can speak for you. Ignore it and it will not go away. Take action and you might have an impact.

Typical Concerns Relative To Volunteer Programs

- Historically – inadequate recruitment efforts
- Increased volunteer turnover
- Government regulations
- They're not trained well enough (competency)
- Lack of supervision of volunteer programs and members
- Lack of public information about volunteer program
- Poor relations between paid and volunteer
- Lack of training for volunteers
- Poor public image
- Lack of quality control
- Lack of availability of volunteers

Planning For The Future

3

Improving Yourself and The Organization

One of the key factors in determining the degree volunteer firefighters shall be utilized within a department is the type and level of services currently being offered. Traditional volunteer firefighters have been used for their suppression capabilities and expertise. In recent years they have also been utilized in providing basic life support within the medical programs offered in many communities. Volunteers are also being utilized in fire prevention, public education, fire investigation, apparatus maintenance, communications, and various support services. From a management point of view probably the best way to analyze the need for volunteer members is to first break the fire department down into systems and sub-systems. By analyzing various services and programs one will be able to determine the capabilities and limitations within an organization. By doing this, for example, you'll be able to determine how many people are available during specific time periods for various activities. Further, you can determine what type of time and staffing requirements would be necessary to conduct fire prevention activities, public education programs, and other activities within the department.

Data can also be utilized to determine response times, fire loss as it relates to response times, location of various fire problems within the community from a historical point of view, and many other valuable input components which are necessary in determining capabilities and limitations both current and future. Once you have analyzed your current delivery system capabilities and evaluated historical data you will be able to answer such questions as: What types of services are the community utilizing? What percentage of calls are medic, false alarms, structure fires, etc.? Who is utilizing the services within the community? When are the emergencies happening? And many other questions which historically have gone unanswered.

Analyzing External Factors

We would be remiss while addressing the need for volunteer firefighters if we did not discuss external factors which might significantly influence the decisions one must make as they relate to fire protection within a community. Community opinions in determining the utilization of paid, volunteer, or a combination of paid and volunteer firefighters are of utmost importance. Politics are a reality within the fire service and must be considered in determining what type of delivery system to utilize. It is the opinion of the authors of this book that a combination paid/volunteer fire department is the most cost effective way to provide fire protection in a small to medium sized community. However, public opinion in your own community will most likely determine the final decision within the jurisdiction. Any major change in

One of the key factors in determining the degree volunteer firefighters shall be utilized within a department is the type and level of services currently being offered.

Politics are a reality within the fire service and must be considered in determining what type of delivery system to utilize.

utilizing personnel to provide fire protection should be carefully thought out and communicated to all individuals involved. A participative approach to planning, organizing and implementing any major changes in a fire department as they relate to staffing is a MUST!

Frequency Of Service Utilization

The frequency of alarms and services utilized by your community is probably the most important factor in recruiting and maintaining volunteer firefighters as it relates to meeting their individual needs. Activity is an important motivator. The "Catch-22" is that if we are doing our job as fire service administrators, we are probably reducing the number of incidents and the amount of loss within our communities, therefore taking away one of the most significant items we have to offer in attracting volunteers. This poses a serious challenge to managers. To overcome such a situation we must look toward external activities which might assist in attracting and maintaining volunteer firefighters. The best substitute for alarm activity has been and remains to be "burn to learns" or training fires. These activities simulate real situations while allowing the opportunity to teach and correct problem areas that may be encountered within firefighting forces. The training and experience which results from burn to learns are very important motivational factors in overcoming the lack of emergency responses. Progressive organizations try to provide simulated exercises, and/or burn to learns, on a regular basis. Two to four burn to learns per year with off months being utilized for simulated exercises such as extrication problems, hazardous materials situations, and various manipulative problems work hand in hand in keeping the morale and attitudes of volunteer personnel up.

A second alternative to dealing with low call volume is to provide comprehensive training for volunteer members. Training time for volunteers is limited, therefore every effort should be made to get the maximum out of each hour devoted to training. Whatever your training schedule, whatever the hours you set aside, is not half as important as the fact that time is safeguarded and every moment during a specific period is utilized to provide quality instruction and training.

Other means of bridging the gap between needs and call volume would be the development and implementation of automatic aid and mutual aid programs, the assignment of the company responsibilities to each member and the establishment of various organizational responsibilities that are important and significant enough to devote valuable volunteer time. Post emergency analysis sessions, or what is commonly referred to as "critiques", provide additional training and exposure to emergency situations also. If handled properly, many times a one or two hour emergency can be utilized to provide up to six to eight hours of instruction and training for members who may not have been involved during the emergency situation itself. This assists in not only improving performance in skill levels but also provides for improved and increased interest.

Conversely, we have those volunteer organizations which respond to a tremendous amount of calls and have an exposure rate that is adverse to long term employment for volunteers, due to too much activity. A volunteer who is called upon more than expected, or is called upon more times then they can afford to respond, will ultimately be faced with a problem just as serious as the volunteer who is not getting

The frequency of alarms and services utilized by your community is probably the most important factor in recruiting and maintaining volunteer firefighters as it relates to meeting their individual needs.

Progressive organizations try to provide simulated exercises, and/or burn to learns, on a regular basis.

enough activity. This situation is commonly referred to as "volunteer burn-out." That is a situation whereby a volunteer is too active and too involved to the point where it is no longer enjoyable. A volunteer who overexposes themselves also risks the chance of causing problems at home or with their chosen job or profession. In these situations, we must assist volunteers in pacing themselves and encourage them to become involved only at a level which is comfortable. One means of avoiding this situation would include but not necessarily be limited to the specific assignment of department responsibilities to balance the work load. The assignment of volunteers to specific companies who are programmed and toned out is also an effective way of pacing a volunteer. It is probably apparent that counseling becomes an important part of dealing with volunteer members in your organization.

Sitting volunteers down individually on a regular basis and discussing their performance and general outlook on their role as a volunteer is an effective means of avoiding problems.

Planning For Fire Protection — Utilizing Volunteers

As it relates to recruiting volunteers, the planning phase basically constitutes a scheme for making, doing, and arranging your resources. The key to managing and planning your volunteer program is to establish responsibility and accountability. For the purpose of this book, we are looking at the planning phase only as it relates to utilizing volunteers. There are three basic levels of fire service planning. The first level is strategic planning, which would include developing a mission statement; vision(s); guiding principles; identifying services and service levels; conducting a "SWOT" analysis (strengths, weaknesses, opportunities, and threats); prioritizing services; identifying customer needs and expectations; and the development of goals, objectives, and implementation strategies. A strategic plan takes a great deal of time to develop, requires technical support, and necessitates a considerable amount of staff time. It is usually difficult for a volunteer organization to support such a complex process without outside assistance. Utilizing corporate volunteers is helpful at this stage of planning. However, the basic concept of planning should be considered when developing your volunteer program. This process would include: identifying the types and level of service currently offered; analyzing current delivery system capabilities; determining services desired as the community grows while building a comprehensive training program.

A second type of plan, as it relates to utilizing volunteer firefighters, would be an operational plan, which would basically consist of a series of plans that would establish goals and objectives from which an organization could function. This plan can be developed internally and updated and reviewed by department members.

This is an excellent means of planning for fire protection and can easily be introduced and implemented in a volunteer fire department. Basically through a participative management approach one can sit down and identify goals and objectives, making sure that they are clear and measurable, while at the same time building ways to monitor and evaluate each specific operation that has been outlined throughout the process. This approach to planning affords excellent ways of making sure personnel are accountable and responsible for various functions and activities. The third type of plan which is necessary within an organization is a tactical plan.

All three levels of planning can be handled through the development of a "customer centered strategic plan."

Customer Centered Strategic Planning For Volunteer Fire Departments

Historically speaking, America's fire service has been trained and well equipped to quickly respond to emergency incidents and evaluate a situation, identify problems, quickly look at alternatives, select a solution, develop strategies, implement tactics and perform tasks in a very quick and efficient manner. Many consider the fire service to be the world's greatest crisis managers. Crisis managers in a positive sense, in as much as we take care of situations very quickly and, in most cases, in an efficient manner. It is only natural that as personnel grow and develop and are promoted within the fire service that they continue to do what they do best, and that is to respond and solve problems. It is, therefore, not surprising that there is not a lot of strategic planning and business planning taking place within the profession.

The customer centered strategic planning process encourages participants to focus on the needs of the local community and the needs of the customers. Businesses employ this type of process to identify market niches. Identifying a niche allows the service provider to focus efforts while reducing risk and wasted effort. The process utilized during the strategic planning process has been developed and adapted to meet the needs of the public service organization. The CCSP process has proven very effective in all types and size of organizations, and in our opinion serves as a model for the fire service. The fire service has entered into a very competitive evolutionary cycle. Public demands continue to increase while dollars shrink, placing even more pressure on the modern fire service manager. Policy makers and organizations need to come up with ways to be more efficient, more effective and in many cases accomplish specific goals, objectives and service demands with less resources.

It would be almost impossible to visit any major private sector company within the country and not have available from them a business plan from which they make decisions, provide focus and direction and communicate both internally and externally present and future expectations of their business. One might ask the question, why wouldn't a public business do the same thing? The answer is emphatically - we should. With this in mind, we will outline a customer centered strategic planning process that has been proven effective and efficient for utilization by volunteer and combination fire departments.

The planning process normally starts with elected officials. As individuals are elected or appointed to oversee a fire department they normally will bring with them their individual visions, agendas, values, politics and opinions. A comprehensive strategic planning process should force the elected officials to consolidate these five areas early on, thus providing the foundation for the strategic plan. The key is to get all of this information on the table and exposed so it can be discussed and ultimately dealt with. One of the most difficult things for any fire department to overcome is a situation whereby individual elected officials have independent agendas which are either not communicated or processed in a collective environment.

Once the individual items listed are incorporated within a planning process they become the basis for the organization's mission, visions and guiding principles.

An established mission, vision statements and guiding principles ultimately become the backbone of the planning process providing direction and focus as well as overall communication both internally and externally.

The mission statement is critical to the planning process for it communicates what the purpose and overall intent of the organization is and in most general terms what it is trying to accomplish.

Vision statements are the organization's forecasting of what they would like the organization to do and look like in the future. A lot of people consider vision statements to be the "dreams" of the organization.

Guiding principles are developed to give overall direction to the organization and its membership as to what the individual and group expectations are and what the parameters are relative to behavior and conduct as well as what the organization perceives to be important and what they value collectively and as individuals as it relates to overall behavior. The best set of guiding principles ever documented and recorded in the history of mankind would be the Ten Commandments.

Once the mission, vision and guiding principles are developed the organization can move on to establishing services, service levels, policies and standards as well as priorities for the organization. These four areas will be dictated by customer needs and customer desires. These needs and desires will be established through communication with the public and other service users within the jurisdiction. Customer needs and customer desires can be established through surveys, questionnaires, interviews and/or random polling of people utilizing either emergency or non-emergency services.

Services must be identified and well thought out to ensure that the organization is providing what it is the public desires with the priority and the emphasis placed on each based on need. Historically, services would include fire suppression, emergency medical services, public education, fire prevention, code enforcement, fire investigation, hazardous material response, urban rescue response as well as a host of other community services such as a chaplain program, critical incident debriefing for organization members, etc.

Identified services are important in determining how resources will be utilized and directed within the organization. But it is also important to specifically identify service levels for each service and program being offered. For example, as it relates to emergency medical services, it is important to determine whether the fire department will provide first response and basic life support or advanced life support and even transportation service from the scene to the local hospital. These decisions will be critical in determining funding and resources necessary to deliver quality service to the community.

Policies and standards must be established by the organization so the membership will know how to perform as agents of the organization. General policies should be developed in key areas including general administration, finance, operations, support services, training, fire prevention, communications and the overall handling of department business. Standards will need to be developed particularly in operations to ensure that the service being provided is consistent and delivered at an acceptable level.

Priorities will need to be developed within the strategic plan so as to ensure that limited resources are being directed and utilized on the most important tasks, projects and functions.

When services, service levels, policies and priorities are established the organization is ready to develop department goals, objectives and performance statements. Overall goals should be established for each major area of the department indicating what is to be accomplished. Measurable objectives will need to be developed in order to provide a method to measure accomplishments as well as to ensure that goals are being met. Performance statements will be developed to provide specific accountability relative to tasks and projects being assigned and accomplished within the organization.

The goals, objectives and performance statements will ultimately become the basis for the evaluation of individual and group performance, compensation and recognition and reward. Progressive fire departments evaluate and measure performance based on the accomplishment of predetermined priorities normally in the form of goals, objectives and performance statements.

The customer centered strategic planning process is a systematic means of managing the overall direction and focus of the organization. The ultimate steps identified within the plan itself would include, but not necessarily be limited to, the development of a mission statement, visions of the future, guiding principles, most important functions and services provided by the department, most critical issues facing the department, most critical customer needs, customer expectations, strengths of the organization, areas needing enhancement, the development of organizational goals, the development of measurable objectives established to meet identified goals, performance statements and implementation strategies.

The strategic plan will become a living document which should be reviewed and referred to on a regular basis in order to ensure that the organization is accomplishing those things that they have identified as important and the department maintains a direction as established by the organization and the community being served. In order to be successful in the future, a comprehensive strategic plan will probably be necessary. If developed and followed, the organization should be efficient, effective and ensure that quality service be delivered to the community served.

Developing Goals And Objectives

The utilization of goals and objectives within a volunteer fire department is becoming more and more widely used as a tool. "Organizational members, whether they be paid or volunteer, have for many years cried out for the following: 'Let me know what you expect of me, give me an opportunity to perform, let me know how I'm getting along, assist me in areas where I need help, and recognize and reward me for my effort." By utilizing goals and objectives as a basis for management a volunteer fire department administrator can meet the above mentions of the work force. Goals and objectives are a basic plan from which all members can perform and function. It is a basic guide which can steer a department through troubled times as well as smooth waters. The purpose of goals and objectives within a volunteer fire department can be summarized into four categories.

1. It gives the organization a sense of direction. By utilizing goals and objectives a department can develop and implement a plan in which all members can see not only where the department is headed, but also what role they specifically play in getting from one point to another. Each individual has a basic need to know what's going on and how they fit in the overall plan. By developing and publishing goals and objectives, a department can meet this basic need.

2. A second purpose for utilizing goals and objectives is to provide an opportunity for all members to participate in achieving pre-determined goals. Each volunteer member has a distinct need to feel they are participating and assisting in the accomplishments of various tasks within the department. In order for this participation to be significant and meaningful, the individual must be challenged and feel they are performing in a manner which is worthy of their participation while challenging and up-grading themselves as a person.

3. A third primary purpose in utilizing goals and objectives within a department is to improve communication. Goals and objectives specifically communicate to the whole organization what the plan is for months and years to come. Any individual within the department should be able to look at the goals and objectives and get a very good overview of what types of activities are scheduled, and to what extent individuals will be involved in activities, during coming weeks and months. Goals and objectives invite members to participate in any portion of the overall plan as outlined in the goals and objectives. Goals and objectives stand as an open invitation for members to participate and become involved.

4. A fourth primary purpose of goals and objectives is to provide a basis for evaluating performance of participating members. By utilizing a goal and an objective plan of action, one will establish a basic foundation from which to evaluate members' performance. As members begin participating in various tasks and projects designed to meet goals and objectives, it will become easier for managers to evaluate involvement, performance, and quality of work within the group. This will give the manager a basis for future assignments as it relates to who is capable of performing, and to what degree they can and will perform.

The advantages of utilizing goals and objectives as a basic management tool within a volunteer fire department are many. Several which stand out and are easily seen within an MBO program are as follows: The MBO system provides for all members to get involved and play a part in the overall plan. It helps to develop a team concept which is essential to us as an emergency response group. Last but not least, it provides a basis for analyzing resources and the effectiveness of the expenditure of various resources in any given area. By analyzing results, as it relates to the number and amount of resources expended, a manager should have a basis for determining cost effectiveness of various programs and projects. A major disadvantage of the management by objective approach to running a volunteer fire department is the fact that it is time consuming and takes a tremendous amount of work in the infinite stages of development.

The utilization of goals and objectives in a volunteer fire department comes in many forms. To try to identify each and every form would be an impossibility. Some of these areas where goals and objectives have been effectively utilized are listed as follows:

a. Reduction of losses

b. Fire prevention activities such as home inspection, school programs, etc.

c. Pre-fire planning

d. Training for results

e. Community projects and programs

f. Improvements of reports and records

g. Recruiting volunteer program

h. Construction of Facilities

i. Modification and/or development of apparatus

j. Development and construction of training facilities

k. Other projects or plans which need to be identified, programmed, monitored and evaluated

Achieving Goals

Improved time management never just happens. Neither does improved personal productivity. Both are results of conscious, deliberate goal setting followed by planning for achievement of goals and taking action to bring goals into reality.

Improved time management never just happens. Neither does improved personal productivity. Both are results of conscious, deliberate goal setting followed by planning for achievement of goals and taking action to bring goals into reality. Goal setting begins with an awareness of the present and a desire to change the future. Since time management is largely a manner of present habits, goals for desired changes must arise from an awareness of the present.

Your Self-Image

Your mental picture of yourself, the image you would like to project to others, and what you truly are all dramatically affect your productivity because your self-image controls how you use time. You act like the kind of person you think you are. It is impossible to act otherwise for any length of time – no matter how much willpower you exercise. People who think of themselves as failures inevitably fail regardless of the amount of time they spend working. What appears to be "trying hard" to succeed may actually be unproductive busy work that reinforces a negative self-image and produces failure. In contrast, people who expect to succeed focus their attention and efforts on constructive activities that produce results.

The attitude you hold today results from a complex history of past experiences and your reaction to them. High on the list of experiences forming your self-image are the opinions and injunctions you frequently heard expressed by parents or other significant adults in childhood. The injunctions most often repeated became part of your thinking and continue to influence your actions. These early teachings were intended to impress upon you the worth of time and the importance of using it productively – for example, "Don't just stand there; do something," or "Don't just sit there; get busy." The subconscious influence of some early conditioning experiences may be positive, but others may lead to an emphasis on external appearances rather than on genuine accomplishments or effective personal productivity.

As a responsible leader, you can choose today to discard any past conditioning that cripples effectiveness and limits productivity. Changing defeatist attitudes or negative thinking is possible because you have infinitely more talent and capability than you have ever used. To strengthen your self-image, begin to identify specific attitudes that have stifled the success you desire and deserve. Set goals to make greater use of your potential. Next, develop a stronger, positive self-image. Desired changes then become a reality.

When you know who you are, you feel secure in the self-confidence that comes from accepting and appreciating your true potential. When you are no longer chained to self-defeating attitudes from the past or fears about what other people might think, you can experience the exhilarating challenge that makes every hour productive. Such confident security is one of the most valuable by-products of a program of personal goals. When you have established worthwhile goals and determined priorities for meeting them, you are free to be yourself – to work and achieve in whatever way is most effective. A positive self-image releases your potential to accomplish remarkable and rewarding results.

Taking Charge

Personal and organizational goals

To gain full mastery of your attitudes, your time and your life, immerse yourself in a total program of personal and organizational goals. Many personal goals involve items money can buy, and your career is the means for earning that money. Other personal goals focus on satisfying such intangible needs as security, ego satisfaction, and self-fulfillment that are inevitably tied to the work environment. When you recognize this relationship both intellectually and emotionally, you realize that productivity leads to the satisfaction of both your personal needs and your professional success.

Reaching your fire department goals requires the cooperation of everyone in the fire department. Ideally, everyone plays an appropriate part in choosing business goals, planning for their achievement, and working out the action steps. Few departments, however, provide ideal environments. Some goals may be handed down to you with little opportunity for your input. You may find it easy to be wholeheartedly committed to the achievement of these goals, but it is also possible that you might find yourself in partial disagreement with a particular goal or plan. At this point, carefully examine your priorities and values to determine exactly how you can contribute to the achievement of stated goals and how you can grow personally by doing so – even though you might have preferred to see the organization move in another direction. Express your ideas about particular organizational goals and plans to the right person at the right time. Only in the case of a serious clash between your personal values and those of the organization will you find it impossible to contribute appropriately. With careful consideration, you can gain insights into ways to contribute to the productivity and long term success of the department.

As a responsible leader, you can choose today to discard any past conditioning that cripples effectiveness and limits productivity.

When you know who you are, you feel secure in the self-confidence that comes from accepting and appreciating your true potential. When you are no longer chained to self-defeating attitudes from the past or fears about what other people might think, you can experience the exhilarating challenge that makes every hour productive.

One element to consider in both personal and organizational goal setting is the time investment required. Most organizations develop more ideas than they have the resources to carry out. Consequently, some criteria must be established for choosing time profitable ventures. Traditionally, these decisions are based on projected return on investment of capital. Obviously, some projects that promise high return require more time on the part of team members than others. In strategic planning, organizations must consider not only the amount of financial involvement but they also must realistically plan for the amount of time required of key people to implement and supervise the project. Some projects that promise a high return on the investment of capital are impractical when the amount of time required by certain team members is considered. This is particularly true when dealing with volunteers.

To ensure adequate time to undertake exciting new projects, all members of the organization need to practice time-proven goal setting principles. This is one strategy that always pays big dividends!

How The Goal Setting Process Works

"Goal setting is the most powerful process available to improve your personal productivity. Without planning and goal setting, all the desire that can be aroused in the limitless potential of the human spirit is wasted like the random lightning of a summer storm." It squanders its force in one flash across the heavens and is lost in the void of space without utility, purpose or direction. It goes unharnessed and unused, its potential power wasted. Ironically, the contrast resulting from its sudden brilliance seems to leave behind an even darker future once the momentary glare fades.

In striking contrast, goal setting – supported by careful planning – provides a sense of direction to keep you focused on the most important activities. Goals serve as a filter to eliminate extraneous demands. Goals bring life to order, meaning and purpose that sustain interest and motivation over a long period of time. Goals evoke your noblest qualities; they express your desire to achieve or to improve your life, and to be more effective, more productive, and more successful tomorrow than you are today.

Although success carries different meanings to different people, there is a definition that fits your dreams as well as that of everyone else.

Success Is The Progressive Realization Of Worthwhile Predetermined Personal Goals

Success does not come by accident; you cannot buy it, inherit it, or even marry into it. Success depends on following a life-long proactive of goal setting and continuous growth – the process of "progressive realization." Success also depends on seeking predetermined goals. Although many worthwhile achievements come as side effects of some other activity or purpose, they are, nevertheless, a direct consequence of the pursuit of predetermined goals. The ultimate effect of reaching a specific goal is not always clearly visible now, but the important point to recognize is that achievement and increased personal productivity invariably arise as a direct consequence of striving toward predetermined goals.

The sole purpose of the goal setting process is to guide you on the entire journey from which to reach fulfillment. The steps in the process are simple but not simplistic, comprehensive but not complex. Be patient and keep an open mind until

Only in the case of a serious clash between your personal values and those of the organization will you find it impossible to contribute appropriately.

"Goal setting is the most powerful process available to improve your personal productivity. Without planning and goal setting, all the desire that can be aroused in the limitless potential of the human spirit is wasted like the random lightning of a summer storm."

the overall pattern of activity begins to unfold. Just remember that you are what you are today because of events that unfolded over time and your choices in response to those events. When you wish to change, to alter attitudes or habits, or to develop new personality traits that will increase your effectiveness, that, too, takes time. Individual pace may vary, but the sequential process of goal setting does not; so follow the plan as outlined. When you internalize the goal-setting process your goals create a magnetic attraction that draws you toward their achievement.

Success is the progressive achievement of worthwhile goals. But before you forge ahead into a complete goal-setting system for your life, you first need to understand and formulate a mission statement for both your personal life and your fire department life because goals grow out of your mission. Mission statements are extremely valuable because what you say you are greatly affects what you actually become. You tend to fulfill your own self-definition; you behave in a manner consistent with the purpose to which you have committed yourself. Possessing a clear purpose and knowing where you are going exert a powerful influence on productivity.

The Power Of Written Goals

A written goals program ensures that you identify achievements that will ultimately prove most meaningful to you. Writing your goals forces you to clarify and crystallize your thinking. A written goals program is also the basis for measuring progress. So commit your plan for personal growth and achievement to writing. Definite plans produce definite results. Indefinite plans, in contrast, produce little or no results. Developing a written plan for achieving your goals provides a number of significant benefits:

Written goals save time

You are continuously bombarded by demands on your time. Write down your goals to keep yourself on course, to minimize interruptions, and to focus your attention. You always know what to do next when your goals are committed to writing.

Written goals help measure progress

Motivation is greatest when there are objectives by which you can measure and monitor accomplishments.

Written goals produce motivation

Goals on paper lend clarity to your purpose and strengthen dedication to their achievement. Written goals remind you of your mission and objectives. Each time you review your goals, you become more excited about working toward them.

Written goals reduce conflict

Conflicts between your values and use of time become obvious when your plans are written out. Written plans help you identify conflicts between various priorities and eliminate damaging frustration.

Written goals form a basis for action

Written goals are just words on pieces of paper until you take action. List specific action steps for moving from the daydream stage to the reality of solid accomplishments. Be sure action steps are logical and practical tasks you are willing and able to undertake. Written plans are the foundation of success, but action is the

Although success carries different meanings to different people, there is a definition that fits your dreams as well as that of everyone else.

Success is the progressive achievement of worthwhile goals. But before you forge ahead into a complete goal-setting system for your life, you first need to understand and formulate a mission statement for both your personal life and your fire department life because goals grow out of your mission.

springboard to actual success and increased productivity.

Written goals stimulate visualization

With your plans written out, you can visualize future results more easily and clearly. You believe more strongly in the possibility of success and become more motivated to reach your goals when you practice the habit of visualization.

Finding Time For Planning And Goal Setting

Powerful timesavers in any undertaking are planning and goal setting. Without them, no amount of activity or hard work ever produces meaningful results or increases your personal productivity. But with them, your efforts propel you toward the progressive realization of your worthwhile, predetermined goals!

The basic challenge in planning and goal setting is finding blocks of uninterrupted time. Interruptions, like meetings, day-to-day routine, and the necessity of dealing with all sorts of major and minor crises, take up time or break it into such small segments that the connected thought essential for effective planning is difficult or even impossible. Remember, most time is wasted not in hours, but in minutes. A bucket with a small hole in the bottom will get just as empty as a bucket that is deliberately kicked over. So, consider all blocks of time – small and large.

With determination, you can find the time you need for planning. Improving your personal productivity depends on it. At the beginning of each week block out specific times to reserve for planning. Mark these on your calendar. Give instructions about how callers are to be handled and what constitutes an emergency worth an interruption. Then follow your plan. An occasional true emergency or unanticipated meeting may alter your schedule. But unless you reserve it and protect it, the time you need for planning will never automatically become available. You do not find time — you schedule it.

As difficult as it sometimes appears to schedule time for planning, a more serious, underlying problem is overcoming the attitudes that frequently stand in the way of reserving time for planning. We are prone to feel uncomfortable unless we are physically "doing" something. We may fear that we are somehow lazy or ineffective unless we are shuffling papers, manipulating objects, or talking about work with other people. We are concerned that someone will catch us sitting apparently idle and conclude that we are "not getting anything done" and have nothing productive to offer the organization. These attitudes are difficult to overcome because they are ingrained by many years of conditioning. But attitudes are merely habits resulting from making repeated choices. You can establish new attitudes and acquire new habits of thought and action by deliberately making new choices, developing a plan for acting upon those choices and taking action on that plan – enthusiastically and persistently.

Tracking And Feedback

High performance leaders are characterized by continuous improvement. Peak performers are also characterized by continuous improvement. A common denominator of continuous improvement of individuals is tracking and feedback. Individual peak performers always use tracking and feedback to improve productivity.

Tracking progress toward the achievement of a predetermined goal provides valuable feedback which enables you to evaluate progress and to make any changes required to reach your goals. Precise, systematic measurement of progress helps you to achieve yet

Remember, most time is wasted not in hours, but in minutes. A bucket with a small hole in the bottom will get just as empty as a bucket that is deliberately kicked over. So, consider all blocks of time – small and large.

Unless you reserve it and protect it, the time you need for planning will never automatically become available. You do not find time – you schedule it.

more progress.

Devising a measuring system also forces you to clarify your goals. Measuring progress may reveal that you need to modify your goals or even that you are working on the wrong goals. Remember, if a goal is worthwhile and is also the right one for you then there are appropriate ways to measure progress toward it.

Tracking progress is the only way to know when you need to take steps to get back on course. Tracking is also the only way to know when you have reached your goal.

Putting Affirmation And Visualization Into Practice

Affirmation and visualization are two tools for improving personal productivity. They transform your thinking, your attitudes, and finally, your behavior. Their impact on attitudes and behavior helps to produce the results you desire.

Affirmation has been given many names – self-motivation, self-commands, auto-suggestion, or self-talk. An affirmation is simply a positive declaration of something you believe to be true or something you expect to become true and desire to live by. The most effective affirmations are those you compose yourself; they are based on your goals and describe the person you want to be, the things you want to do, and what you want to possess. When you repeat such affirmations, you build the needed internal confidence and determination to overcome obstacles, accomplish goals, and improve productivity.

Use affirmations as positive material for your mind to act upon in building constructive attitudes. For example, if you want to gain greater control over your emotions, remind yourself daily, "I control my emotions and reactions at all times. No matter what other people say, think, or do, and no matter what circumstances arise, I remain calm and in control." Another example dealing with communication could be, "I enjoy knowing the people in my fire department as individuals. I listen when they talk and understand both their words and their feelings. I respect them and their right to be themselves."

The mind is like a highly efficient computer. It controls emotions, attitudes and actions according to the information it has been given to work with. If you feed your mind negative ideas, it can only respond negatively. But when you give it constructive, confident directives, it responds with positive motivation for productive action.

> **Organization Goals and Objectives**
>
> Give the department direction
>
> Provide members an opportunity to participate
>
> Improve communications
>
> Provide a basis to evaluate performance

A second useful technique for focusing your creative power on your goals is the practice of visualization – the force that transforms your dreams into reality. Visualization is the act of mentally picturing ideas, events, circumstances, and concrete objects. The importance of visualization in goal setting is its effectiveness in enhancing your ability to achieve.

Visualization is exercised by successful, high achievers in every profession. The visualizer

Individual peak performers always use tracking and feedback to improve productivity.

Tracking progress is the only way to know when you need to take steps to get back on course.

clearly and distinctly "sees" the results that will come from the persistent pursuit of goals. When you see a vivid picture of yourself in possession of your goals, the picture stimulates desire, sparks creativity in planning action steps, and fuels motivation to take action. In the majority of situations, vision gives more accurate knowledge than any other sense. This truth is reflected in the fact that we customarily think in pictures; in other words, we visualize.

You can, by conscious practice, refine your skill in visualization and turn it into a powerful, forceful habit that improves your personal productivity. Practice the creative ability to visualize, and support your visualization regularly and according to a plan. You will find it one of the most helpful tools you have ever used for harnessing the power of your imagination. Visualization affects every part of the goal-setting process.

Visualization Focuses Your Attention On Your Goals

You achieve a goal only when you know exactly what it is you want. Visualization is the tool that brings a goal into sharp focus so you take only actions that move you in the right direction.

Visualization increases desire

When, through visualization, you experience how it feels to be in possession of your goals, desire grows by leaps and bounds. Without desire, there is no life nor excitement in your goals program. Enthusiastic desire sustains motivation throughout the entire process of setting and achieving goals.

Visualization intensifies beliefs and commitment

The saying, "seeing is believing," has more than just a grain of truth. When you visualize yourself in possession of a goal, you believe in your ability to achieve it. You know what it looks like, how it feels, and what you must gain in the way of knowledge and skill to possess it.

Visualization sharpens concentration

Because visualization shows you the exact path to your destination, you are not distracted by outside circumstances or the urging of others to leave the path you have selected. You move directly toward the chosen goal.

Visualization relieves stress

Anxiety and stress creep in when doubt, uncertainty, and fear are associated with the future. Visualization prevents and relieves stress by providing believable information about the future.

Visualization fuels motivation

Visualization generates intense interest and a sense of urgency that keep motivation at a white-hot heat. Procession, inertia, and indecision disappear. You are energized and eager to keep moving toward the accomplishment of your goals.

"Take charge of your life. You can do with it what you will," said the Greek philosopher Plato. These words are still true today. You can do with your life whatever you will when you make the most of every minute. Take responsibility for your productivity by managing your time more effectively. You will be astonished at the results.

Whatever you vividly imagine, ardently desire, sincerely believe and enthusiastically act upon must inevitably come to pass. You and your fire department will ultimately benefit from it.

Why Do People Volunteer?

4

Volunteer Fire Departments Provide Basic Needs

Every person has the need to feel important, accomplish important things, and to grow as an individual. The volunteer fire department, if managed properly, provides for these basic needs.

If we were to analyze the average person who we will be recruiting throughout the community, we would find that most of the basic needs which are so important to that person aren't being met through their employment. Therefore, the individual will look outside the workplace for some activity or relationship which will assist them in meeting many needs they have as human beings. Some of these needs would include:

- a sense of belonging
- achievement
- economic security
- increased responsibility
- self-respect
- understanding
- challenge
- recognition
- reward
- growth and development
- need to have fun and enjoy one's environment

If we were to analyze the ability of the workplace to provide for these needs, it would be quite obvious that a person doesn't come close to meeting basic needs on the job. To the contrary, if we were to analyze the fire department's ability to meet these needs, we would find we have a tremendous amount to offer volunteers. By the nature of the services provided, and the type of atmosphere in which it's provided, it's no wonder that departments who are doing a quality job find it easier, rather than harder, to attract volunteers.

You will be given many reasons when you ask new members why they joined. These will include:

- helping the community
- helping a fellow person in need
- possibly saving a life
- to someday become a paid firefighter
- my neighbor is one
- I had a fire once and want to help protect others from that sort of devastation, etc.

The list goes on and on. Contrary to all these explanations, the primary reason people volunteer has always been, and will remain to be, "the desire of the individual to meet their personal human needs." Once you understand why it is people come to your department and want to join, the easier it becomes to develop programs that will not only attract them, but will keep them for years to come. Meet the needs of your membership and you will minimize volunteer turnover.

Years ago departments thought the way to recruit and maintain a volunteer force was to wine and dine them, while providing them with a very social orientated environment. Through a considerable amount of research and analysis, it has been found that not only is this method of attracting and maintaining volunteers a poor one, it is, in most cases, nonproductive. The way to recruit and maintain proficient volunteers is to meet their needs and provide them with the chance to deliver a quality product to the community. That product, in our case, is "service." This is accomplished by developing programs, including a progressive, compressive training program.

Departments which seem to experience the most success in recruiting and maintaining volunteers are those requiring and demanding a great deal from their membership, while at the same time providing them with a chance to participate and grow as individuals. A program which is faltering or that finds it more and more difficult to compete should first look at the quality of the product and second look to see if they are meeting the needs of their volunteers. Once a department finds itself in such a situation, it becomes necessary to make some significant changes. As is the case with most change in the fire service, it must be carefully planned out, be communicated to all concerned, and it must encourage participation from all members.

It is not a sign of weakness to admit there have been mistakes made or to admit shortcomings. To the contrary. To bring these items out in the open and request assistance in getting the department back on track is a sign of a person who not only cares, but is willing to put their personal feelings aside for the best interest of the group. This approach will be appreciated and provide a basis for others to follow.

Who Is Responsible For Recruiting?

The question, "Who is responsible for recruiting?" is asked time and time again. The answer is simple: every member of the fire department. Every member from the Fire Chief down must share the responsibility for recruiting new members.

Recruiting starts with the Fire Chief. They are the organization's super salesman. The recruiting of new members should be an intricate part of the overall goals and objectives developed by the department's management team. Members must see that the overall plan is carried out.

Recruitment of volunteer firefighters is "sales." As a member of the fire department, everyone will be responsible for selling the product, just as a vacuum cleaner salesman is responsible for selling their product.

The criteria for success is also similar to that of the vacuum cleaner salesman. You must first have a product that is sellable, one in which every member can honestly be proud to sell, and everyone must be able to express genuine enthusiasm towards the product being sold.

In the case of the fire department, that enthusiasm and pride are simply feelings toward the fire department, as represented by every member.

As mentioned earlier, the Fire Chief must be the super salesman. They must lead the way in the recruitment effort. If recruiting is to be taken seriously by department members, it must be taken seriously by the Fire Chief. The Fire Chief should be available to support and confirm statements made during recruitment efforts. Each and every person within the organization must believe in their fire department and must believe in their volunteer force.

Who Are You Competing With?

As mentioned earlier, recruitment is sales. To successfully sell a product, you must know whom you are competing with. You must know what they are selling and why people might be attracted to their products. Hopefully, if your department has done its job in organizing and planning for volunteer firefighters and your product (the fire department) is a good one, you will be successful in outselling your competition. This would include providing a good recruiting program, a progressive training program and maintaining an atmosphere which is conducive to keeping volunteers. Once this preparation work is done, you should be in a good position to take on the competition and be successful.

As strange as it might seem, your competition doesn't come from other fire departments. It's usually other activities, events and organizations which community members have available to them, which are normally occurring during the same time periods when you would be needing them as volunteers. The only way to sell them on your program is to offer them more than the competition can offer.

There's no easy way to sell and keep volunteers. There are no shortcuts to success. Bowling leagues, television, civic organizations and various local activities all stand as potential competitors. All of these items mentioned can either complement a volunteer program or steal members from it, depending on the quality of your fire department and its ability to attract and maintain membership. Again, the key to your success is to have a product that out-sells the competitors.

What Do You Have To Offer New Members?

The most common question asked by potential volunteer members is, "Why should I join?" The answer is quite complex and will be answered in detail when we discuss how to maintain volunteers.

At this point we can state, however, that the primary reason the individual has turned to the fire department is to meet their personal needs. Therefore, it is safe to say if we are providing the type of programs necessary to keep volunteer members, the answer to their question is, "You should join because we will meet your personal needs as well as the needs of the department and community."

You are in competition for volunteers. You must know what your competition is selling and why people might be attracted to their product(s).

Fifty years ago the number one recruiter in the world used a simple picture of Uncle Sam with a statement underneath that read, "I WANT YOU!" We assure you that the young people today would have difficulty relating and responding to that campaign.

Recruitment Techniques – How Times Have Changed

Fifty years ago the number one recruiter in the world used a simple picture of Uncle Sam with a statement underneath that read, "I WANT YOU!" We assure you that the young people today would have difficulty relating and responding to that campaign. There is little doubt that techniques and approaches must change to attract potential volunteers today.

How Do You Sell Your Fire Department To Potential Volunteers?

As mentioned earlier, you sell your department through offering a quality product. You must also project a positive, enthusiastic attitude toward the department and the activities which it offers the community and its membership. People are looking for something better than what they have, or have experienced. They're also looking for something different than what they are accustomed to. The fire department, if it's an active one, should be able to meet the expectations of the new member.

We firmly believe that once programs are developed and quality is built in, the department will in most instances sell itself. By being out in the community and being actively involved in important functions on a daily basis and providing quality service in the areas determined to be important in your jurisdiction, you should find that the work you do sells the department. Several very effective means of assisting a department in attracting potential volunteer firefighters would include the development of recruitment flyers, a systematic means of distributing literature and information, and the development of visual programs that can be used to emphasize certain aspects of the organization.

Many departments have produced excellent slide programs which illustrate various activities within the department. The old saying, "One picture is worth a thousand words," is definitely true when one is trying to illustrate the various functions within a fire department. These items, in conjunction with the need for the recruit to decide whether or not the fire department is right for them, provides a basis for recruiting new members. There are a lot of people throughout the community who have expertise in these areas and who would probably be more than willing to assist your department with various aspects of the recruiting phase. High schools and junior colleges both have the resources to assist in the development and distribution of recruiting information and literature. Probably the best recruitment tool you have at your disposal is the current members of your department. Approximately 50 percent of all volunteers are recruited by other volunteers.

Profile of Volunteer
20 – 39 years old
80% are married
96.7% male
45% blue collar worker
50% have friends as volunteers
73.1% volunteer for the fire department only

What Type Of Person Are You Looking For?

For the immediate need within the fire department, there is no question that you should be looking for blue collar workers between the ages of nineteen and thirty-five years. Historical data has proven there is an excellent chance that high school students won't stay with you once they graduate from high school. There are several things that usually happen to high school graduates. First, they will graduate and go off to college, and therefore won't be available; they may take a job in another community and therefore won't be available; they may join the service making them unavailable; and, last but not least, they may settle within the community in which they graduated. As you can see, the odds are against you maintaining that individual as a volunteer firefighter.

We aren't telling you not to use high school students, however we feel the risks are great that the individual will not ultimately be productive in the department and therefore the cost effectiveness of training and trying to maintain those types of individuals is probably not in the best interest of the department. We feel the explorer post and cadet programs are cost effective; you're using the young people within the community and providing them with specific training, but you're not depending on them to the point where the program is in jeopardy if a group of them terminate. Therefore, young people between the ages of fifteen and eighteen years of age should be considered low priority and dealt with in a low-key manner.

A third area of consideration would be that of specialized recruits. There are many functions within the department requiring personnel that don't mandate they be in good physical condition and available throughout the week. We classify these individuals as "specialized recruits."

Fire prevention activities, EMS programs, public education, data entry and analysis, clerical work, and maintenance are all areas which can use specialized recruits. These recruits might  include the elderly, the handicapped and those members of the community who physically aren't capable of fighting fires.

One can analyze various programs within a specific department and determine if, in fact, specialized recruits could assist in meeting pre-determined goals and objectives within a department. By utilizing specialized recruits, much of the time burden is taken off your main firefighting force. As stated previously, to over-work a volunteer and require more than they are able and/or expected to give, could cause premature burn-out of an important person or groups of persons within the department.

The most under used volunteer resource in most communities are the young and the old...both groups have a lot to offer.

In reviewing what type of person you are looking for, again we find that the blue collar worker is our prime target. Young people are the target for the future needs within the department and specialized recruits are the target for assisting in various support functions.

Character Qualities

Some of the character qualities to look for in potential recruits are:

Leadership experience. This is highly desirable, and often indicates whether the individual has officer potential.

Maturity. Can this person take constructive criticism? Will they be able to accept discipline and commands?

Commitment. A recruiter may ask about organizations the person has belonged to, and how long have they stayed with them. This gives some idea of the likelihood of commitment and tenure. People who tend to leave organizations after serving a short time should be questioned about the likelihood of a long-term commitment to the fire service. (Turnover is expensive.)

Team player. This is important because fire fighting involves teamwork.

Conscientiousness. A dedicated, conscientious recruit will be thorough in both operational and administrative duties.

Interpersonal skills. Can this person relate well with others? Are they a good communicator? Will they fit in?

Moral character. Can they be trusted in members' homes and businesses and around the station? Have they experienced difficulties in the past?

Problem solving ability. People who can solve problems will help work out solutions and not just complain when a problem arises.

Initiative. This may be a key indicator of how involved and how much of a leader a recruit will be.

Recruits should be asked what they expect from the fire service. In turn, they should be told what is expected from them. This may weed out some people immediately before going through the lengthy screening process.

When Do You Recruit Volunteers?

Volunteers are recruited by the departments that take the recruitment functions seriously every day of the year. Volunteers are recruited at parties, in the barber shop, on the fire scene, at PTA meetings and every other event or activity where people gather. Each member of the volunteer fire department should always be ready and prepared to get names and phone numbers of potential volunteer firefighters. Many departments provide members with business cards to be used in gathering information and making a link between a potential volunteer and the fire department.

If someone in the community shows an interest in your fire department each and every member of that department should be briefed and trained as to how to get that person to the point where a presentation can be made and a possible contract can be signed. Community service groups throughout your area can work as exceptional recruiting centers. People who are actively involved in service groups

Each member of the volunteer fire department should always be ready and prepared to get names and phone numbers of potential volunteer firefighters.

such as the Chamber of Commerce, the Elks, the Eagles, Rotary Club, etc., have already demonstrated their desire and ability to get involved within the community. Therefore, they would make prime targets for the fire department.

Members of the department can also recruit new members at their place of employment. Teach your personnel to discuss the fire department, while being prepared to answer any questions potential volunteers might have.

Once members of a department get accustomed to gathering names and phone numbers in order to assist in maintaining a potential volunteer list, the designated volunteer coordinator can proceed with the recruiting and screening process.

After names and numbers are gathered, the next step would be making contact with the potential volunteer. A telephone call usually works quite well for an initial contact. When you call a potential volunteer whose name and number you've been given, keep the conversation brief and try to get an appointment with that individual. You should first ask if they are still interested in becoming a volunteer; if the reply is yes, find out when and where you could meet to give them more information regarding the fire department. At this point, hopefully an appointment will have been made and you'll be able to provide a presentation to the person stating the services offered, what the department will do for the individual and what he or she can expect from the organization.

Once the appointment is made and the program is presented, the person should be kept on an eligibility list until such time as the department is ready to run a recruit academy. We discourage any organization from signing up volunteers and putting them on the roster prior to running them through a comprehensive recruit program. Remember, "The key to your success in maintaining volunteer firefighters is to provide a quality program." In order to provide a quality program, you must fill your ranks with quality people.

When the eligibility list is developed, the recruiter should keep in touch with every member on the list to let them know when and where they can expect to be utilized. Once the eligibility list contains the number of individuals desired, a recruit academy can be scheduled.

Keep in mind that most departments running an intense recruit program from forty to one-hundred sixty hours usually experience about a 50 percent drop-out rate. In other words, if you're anticipating the need for ten new volunteers, you would likely recruit and place twenty individuals in the recruit training program. They will reduce in number for various reasons as the training program continues. The advantage of recruiting and expecting a 50 percent return is quite obvious. You'll be receiving the "cream of the crop" which will tend to strengthen your existing volunteer force. Those that shouldn't be there will drop out for various reasons.

Where Do You Find Volunteer Firefighters?

Volunteer firefighters come from all walks of life. Remember, the majority of your force should be made up of blue collar workers. Therefore the majority of your recruiting efforts should be in

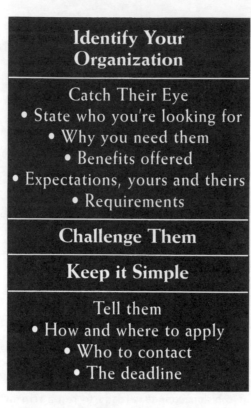

Identify Your Organization

Catch Their Eye
- State who you're looking for
  - Why you need them
  - Benefits offered
- Expectations, yours and theirs
  - Requirements

Challenge Them

Keep it Simple

Tell them
- How and where to apply
  - Who to contact
  - The deadline

We help you, help us to help others

Help your community and become someone special!

Learn the skills necessary to protect yourself and families in an emergency situation. Become a member of a Community Action Team!

If you are between the ages of 16 and 50 and are
- In good health
- Are a resident of the Fire District
- Can meet the training and attendance requirements

Your Fire Department is willing to train you for volunteer openings in Firefighting, Emergency Medical Technicians, Apparatus Operators and other positions

No experience is necessary. All equipment and training will be furnished. The next training program will begin in May of 1999. Call 555-5555 or better still, come in and talk with one of the representatives at the Downtown Fire station.

plants, corporations and those areas throughout the community which employ workers who are accustomed to "getting their hands dirty."

Probably the best resource available in recruiting volunteers is other volunteer members themselves. Friends of existing volunteer firefighters have proven to make excellent additions to departments. Usually existing volunteers will only recommend those friends or acquaintances in which they themselves are quite confident. They know their name and reputation is somewhat on the line when they bring a new member into the department, so they're usually quite selective in suggesting and recommending individuals. Consequently, the chances of success are significantly increased.

Many organizations provide prizes and/or rewards for those members who bring in the most accepted volunteers within a given time period. This type of program has many merits, but should also be monitored quite closely to make sure you are receiving quality and not just quantity.

Booths in public places, participating in community activities and making periodical visits to commercial/industrial properties throughout your community are all quite good means of gathering names and phone numbers of potential volunteer firefighters.

Flyers play an important part in the recruitment of volunteer firefighters. A comprehensive recruitment flyer would follow these guidelines: identify your organization; catch their eye (by stating who you're looking for); tell them why you need them, benefits offered, expectations (yours and theirs), requirements; and, last but not least, provide them with information on how to apply, where to apply, who to contact and the deadline. Once the person uses a format as outlined, the best advice is to keep the flyer simple so it will be read.

Radio, television and newspapers can be excellent means of conveying the need for volunteer firefighters.

How Do You Get Them To Volunteer?

There are many reasons why volunteers want to participate in a fire department. As stated earlier, the primary reason is to meet their needs, so the best way to get persons within a community to volunteer is to focus on meeting their needs

Challenging them is the quickest way to meet their needs. First, let them know it's not going to be easy to become a member of your department. The success rate of those

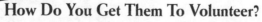

persons who have tried is quite low. The simple reason is you're looking only for quality membership. This in itself will probably arouse the curiosity of the individual and will surely provide an atmosphere whereby they'll want to prove they can meet your standards.

Other means of challenging them are by putting them through a very comprehensive recruit program that requires both physical and mental demonstration of their abilities. A recruit program should be difficult, intense, and therefore rewarding, once a person succeeds in completing all requirements.

Another method of motivating people to volunteer is to show them results. Show them what your department and your programs have done for other volunteer members. Show slides and photos; have members give testimonials citing how good the department is and what it can do for the individual; show the reward, recognition and satisfaction they'll receive if they become a part of the organization.

The key to getting people to volunteer is to have a quality product, be proud of that product, and meet people's needs.

The Roles Of The Volunteer Coordinator

Every volunteer fire department, no matter how big or small, needs to assign a person to be the volunteer coordinator. The volunteer coordinator has the responsibility to plan, organize and coordinate the recruiting, training, and maintenance of the volunteer force. Responsibilities might include:

- plan and schedule activities
- determine the roles of volunteers
- coordinate recruitment efforts
- keep records and write reports
- write job descriptions
- prepare the volunteer budget
- interview and screen potential members
- performance review
- evaluate future needs
- public relations and information
- counsel volunteers
- testing and measuring
- orientation programs
- liaison with staff and administration
- make formal presentations
- recruit additional members

Three-hundred and sixty plus years ago, when the first volunteer fire departments were organized in the United States, they were formed on the premise of four factors being present: That all volunteer companies would be proud, exclusive, influential, and competitive.

I maintain that if we are doing the job in our departments today, they should still be very proud, exclusive, influential, and competitive. To do any less than what is required to meet these four criteria is to shortchange yourself, your department, your volunteers and your community.

Three-hundred and sixty plus years ago, when the first volunteer fire departments were organized in the United States, they were formed on the premise of four factors being present: That all volunteer companies would be proud, exclusive, influential, and competitive.

You must continually strive toward excellence in providing the best levels of service no matter what the situation; we must meet these four requirements.

There are four basic ways to market

Advertising

Publicity

Promoting

Personal Selling

Marketing Your Volunteer Fire Department To Potential Volunteers

Marketing of the fire service is based primarily upon perception. Perception in most cases is based upon image. People perceive value if they receive service when they need it and if that service is what they feel they need.

Trust is the cornerstone of relationships. What we're really talking about when we discuss marketing is building relationships. We count on these relationships to sustain our departments.

Marketing is important to the overall success of your VOLUNTEER FIRE DEPARTMENT. There are many things that go into marketing but there are four basic ways to market: advertising, publicity, promoting, and personal selling.

The fire service needs to understand marketing and become more involved in marketing the fire department to the public. Marketing is the process that will help leaders of the volunteer fire department justify the things the department needs. Needs directly relate to marketing. The public is the market that you must work to convince that the things the fire department wants are justifiable based upon the return of value that the public perceives it will receive.

Marketing Is The Caring Trade Of Value For Value

This means that when our fire/rescue organization identifies a need it has for volunteers, it figures out what value it might offer those volunteers in exchange for their efforts.

The true value of marketing comes when all parties involved in an exchange relationship are convinced they have received the greatest value. People can relate to value only when they can feel the worth. In EMS this may be hard to do, but it is not impossible. One way people feel value is when they receive service, but another way for value to be perceived is through an effective public relations program that involves being out with the public where volunteers do things (shopping malls, sporting events, social events, etc.).

Like any of the tools needed to manage a well-run program, it takes a lot of hard work, creative problem solving, and innovative thinking to make it happen.

Everyone can do it.

The Importance Of Marketing

Why marketing is so important:

1. **In recruitment**: This is the bridge between potential volunteers and your program. It defines what is needed and what benefits are available in addition to offering incentives to prospective volunteers in order to attract and involve them in your program.

2. **In retention**: The values given volunteers in the way of tangible and intangible rewards are what keep volunteers coming back and also stimulates them to tell others about their experiences, which feeds back into recruitment.

3. **For organizational climate**: Happy workers who feel good about their work create a climate that stimulates creativity, enjoyment, and achievement. It simply "feels good" to work and be there! This is catching!

4. **In gaining support**: When as a Fire Chief, you need the support of administration, board members, volunteers, peer level department heads, etc., marketing provides the basic leverage that allows you to define an exchange relationship (what's in it for them and you) to attain your goals.

5. **In winning public support**: As you need support of the general public for your programs, events, philosophies, etc., marketing again holds the key to attaining it. The arm of public relations comes into play in this relationship.

6. **In resources development**: When you are in need of goods, services, or dollars, the marketing orientation is the magic that will produce results. In fact, the use of marketing holds the key to getting what you need without spending dollars... the art of fund raising.

7. **In obtaining members/participants/consumers**: Marketing is again the key in getting people to accept your services and/or products. Its proper application convinces people to become involved with you as they perceive a great value to themselves for that involvement.

The public in most cases does not know who their fire department is. They do know that if they have a need they can pick up the telephone and dial 9-1-1. 9-1-1 is generic, they don't know who they get when they call that number, but they do know they will get help. And as the vice president of Intel once said, "How dare a fire department be anything but the best when the person calling has no choice in who provides the service."

The public perceives the red fire truck as the symbol of their fire department. The name on the side of the fire truck is not important to most people. They just want a fire truck when they need it.

Too often marketing is the missing tool in the Fire Chief's toolbox with the result being a deficiency in recruitment, retention, and fundraising. As shrinking resources, higher demands on volunteers' time, and greater demands from the public impact us more dramatically each year, we must master this tool in order to achieve the successful attainment of our goals.

Marketing is made up of three components: Publics, Markets, and the Exchange relationship. Let's define all three.

Public: Any identifiable segment of your community that surrounds your program or agency. Publics are groups that you might or might never want to have a relationship with. When identifying publics you don't worry about or try to judge whether or not you might someday want to interact with them...they simply exist.

Stakeholders are another form of public. The public must realize that they have a stake in the successes of the fire department. In order for the fire department to be successful in delivering services when the public needs it, the public must have invested in the department a long time prior to needing the services of the department.

Market is an identified public with whom you decide you wish to establish a trade relationship. To put it in simpler terms: They have what you need/want.

Marketing is made up of three components: Publics, Markets, and the Exchange relationship.

Why should the public feel safer just because they have you for their fire department? Who is your target audience? In your community everyone receives service from the fire department. That's one of the many challenges facing the Fire Chief. Each fire department develops a persona within that organization. That persona can transform to the perception that the public has of the fire department.

Market: Market is an identified public with whom you decide you wish to establish a trade relationship. To put it in simpler terms: They have what you need/want.

Of a list of 100 publics, you may find a half dozen that could meet your need if they choose to do so. The trick is having a list of publics from which to choose. Without it, groups suffer from "resource myopia," never realizing how many potential resources are available to them.

The fire department with its multiple targets is very similar to McDonald's. The marketing effort must simultaneously talk to each demographic group individually. When we talk to each demographic group we have to say essentially, "We understand you, we understand your lifestyle, we understand your culture, and you're important to us." The message that the department develops has to talk to the targeted group. No one program will be successful with each demographic group within the community. To be successful in marketing the department, its services, and issues (fire prevention, public education), you have to develop several messages for each group. Again, look to McDonald's; they market to every demographic group that is out there. You see tremendous diversity in their marketing campaigns. The primary target markets for McDonald's include: kids and families, teens, young adults, minorities and seniors.

The fire department in most communities can develop programs for each demographic group listed by McDonald's and be very successful.

Exchange Relationship: This is the key stone of success, it is the bargain that is struck between the volunteer fire department and the markets who have what you desire. It is the essence of marketing – the trade of value for value. Its qualities include:

1. Honesty and fairness

2. No hidden agendas or pitfalls

3. A user-oriented position

4. A targeted approach

5. Highest concern for what the other party will receive as value

6. Attention to your success in attaining goals

7. A lot of homework!

When thinking about marketing your volunteer fire department, it is vital that you understand the four step process that will bring success.

The four steps are:

1. What do you have?

2. What do you need?

3. Who has what you need?

4. How do you get what you need?

In most instances, people begin with step #2 – their needs – then find themselves with many problems, including resource myopia.

What Business Are You In?

When this subject comes up in the fire stations the most common answer usually is: "We're in the fire business!" The fire business must be defined within each organization. The reality of today's definition is that the fire business extends dramatically beyond simple fire suppression actions. You must define your business prior to developing your marketing program. It is extremely important that your scope of operations be specifically defined. The fire department and its members have limits. Limits relate to training, education, skill, and experience of the members. The firefighters within each organization cannot in most cases be experts in every type of emergency that the public may encounter within their lives. Maybe you're not in the fire business?

What's Your Product?

The major product that we offer the public is "service." Service is defined as an act of serving others. The fire department product is customer service. The community has a right to expect certain things from the fire department. But the fire department has the ability to define its product. Today, there are so many services that the public may perceive they need but there are limits to money that can impact the potential for delivering those services. In the current political climate and budget constraints different methods of delivering services must be explored by the Fire Chief. You're in the service business!

Three Product Classifications

Marketing products are classified in three ways that have great significance for the fire department when translated to our terminology.

The expected product is defined as a customer's minimum expectations concerning price, delivery and technical support.

In a volunteer department this can be translated into the expectations volunteers have when they offer their energies to work with you:

1. **Price** — What it will "cost" them in out of pocket expenses, energy and emotional drain.

2. **Delivery** — When, where, and how they expect they will have to perform their duties.

3. **Technical support** – How they will be trained to carry out assignments, who will help them, who will oversee their efforts, how the other staff, especially paid, will work with them.

The Augmented Product is defined as offering the customer more than they expected or think they need.

In recruiting and retention issues this can be how a volunteer perceives that they are exceeding their minimum expectations by receiving the added value of knowing they are truly making a difference, building skills, feeling needed, having their efforts recognized and/or simply having fun.

The wise Fire Chief keeps a close tab on this augmentation, being aware of the dozens of "benefits" that might be offered to volunteers in order that they feel good about their work, tell others, and stay longer.

The only "danger" in augmentation is that these additional benefits may begin to be regarded as normal, thus setting up higher expectations for future efforts.

The Potential Product

What business does the public think you're in? It is important that you assess perceptions of people outside your immediate circle to test the accuracy of their perceptions as to what business you are in and what products you offer.

What products do they think you offer?

Advertising is a campaign that is designed, pre-tested on target markets, post-tested for results, and presented through selected media in order to influence people to accept a product or idea. It usually has a theme or slogan and is targeted to specific audiences.

Too frequently, volunteer fire departments misunderstand the purpose of advertising and therefore they omit testing, targeting, and life-cycle checks, making their ads ineffective.

Advertising can serve a fire/rescue organization by making people aware of what they do, what they need, how to contact them, etc. It is usually designed for "quick consumption," not lengthy reading. It is designed to stimulate familiarity.

Tragically, many people confuse advertising with marketing, telling me they have a "wonderful recruitment campaign" going on – twenty-four ads in papers and one-hundred posters in stores. This is not a political campaign – it is an advertising effort that can augment recruitment through personalized asking, but by itself is not going to be successful in acquiring volunteers, donors, etc.

The most remembered and successful advertising of this century was the "UNCLE SAM WANTS YOU!" campaign that appeared in every post office and train station across the USA in the 1940's.

Had the posters and ads, however, not been backed up by recruitment offices, appeals, draft, and the fervor of people wanting to help their country, the ads themselves would have fallen short of their goal.

Publicity is the development and dissemination of news and promotional material designed to bring favorable attention to a product, person, organization, place or cause. Publicity is also cost effective. It is usually free, but requires an active Public Information Officer. Publicity can also be more credible than advertising.

- It is different from advertising in that it is usually carried by the media free of charge and without indication of the source.

- Publicity is a creative challenge for your group, as you find ways to use media to tell your story and express needs for support.

- It is critical for you to check the effectiveness of your publicity by finding out how well the public is getting your message.

- It is publicity that plays a key role in perceptions and needs to be current, clear, and realistic as it comes across to the public.

Promotion is defined as marketing activities other than personal selling and the two mentioned previously that project your message. Such efforts might include displays, speakers bureaus presentations, booths at local fairs and events, etc. These efforts are usually unique and are not necessarily expected to occur again.

Again, promotions can stimulate the creative minds of your people, as they think of ways to positively gain attention and share information with the public.

Personal selling: The effective Fire Chief understands that some of the best recruiters for their volunteer fire department are satisfied volunteers who relate their positive experiences and encourage others to become involved. Satisfied donors and supporters can and do play the same role reaching out to your story.

The "catch" in understanding personal selling, however, comes when you separate this promotional activity into its two categories: 1) spontaneous and 2) targeted.

Marketing may be a department responsibility, but is everyone's job to some extent, just as recruitment is everyone's job.

For the most part, the contacts made by volunteers and supporters is sporadic, unplanned and happenstance. A citizen is the most important person to come in contact with your fire department, in person, by mail or by the telephone. A citizen does not cause an interruption of our work. A citizen is the purpose for our existence. We are not doing the public a favor by serving. A citizen is not an outsider to our activities. A citizen is not someone to argue with or match wits with. Nobody has ever won an argument with a citizen. This is spontaneous personal selling, and it is difficult to measure or control. Most of the people involved in this form of promotion do not even realize what it is; they are simply sharing their pleasure with others.

Volunteers must be trained in developing and maintaining the fire department image with the public. Volunteers must show a genuine interest in the citizens' problems. How volunteers act, speak, look will have some impact on the public's perception of the volunteer personally and the department as a whole. This issue can create a ticklish situation for the leadership of the organization. Volunteers in many cases don't realize that they are public employees. In many cases because of insurance requirements it is in the state law. As a public employee the volunteer must understand the trust the public has put in that individual as well as that organization. The way a volunteer acts, speaks and the appearance of the volunteer could have a negative impact on the public perception of the organization.

Although not always permitted, it is important to realize that when a volunteer puts a light on the outside of their personal vehicle, the public perceives in many cases that the vehicle is an official vehicle. The way a volunteer drives, parks or even the location of the vehicle could have an impact positive or negative on the public perception of the department.

The name of the department is a precious commodity and must be guarded. When you look at McDonald's sign – what do you notice about the sign? It is clean, illuminated, and highly visible. The same goes for the department name. Wherever that name is allowed to be placed could change the public perception of the organization it represents.

In the volunteer fire service t-shirts, jackets, etc. are marketing tools. The impression that this clothing leaves can produce positive or negative effects.

Leaders within the volunteer fire department need to set a good example by training their personnel in dealing with casual contacts through neighbors, relatives, storekeepers, church members, and local organizations. People will judge the fire department by the few people they know who volunteer for the fire department.

The way a volunteer acts, speaks and the appearance of the volunteer could have a negative impact on the public perception of the organization.

Are your people trained in public relations? Do your members understand the importance of public support to the fire department? Before allowing members to meet with the public there needs to be some type of training/ education.

The volunteer who criticizes their department, who gripes about conditions, spreads stories of waste and inefficiency, can do much to harm the department.

The volunteer who is disorderly, dirty or sloppy discredits the whole organization. So does the one who never pays their debts or who otherwise violates the rules of good social conduct.

The second type of personal selling, targeted, involves people who are interested in participating in events where the public is expected to attend. This type of selling is planned. The volunteers involved in targeted selling could even be coached as to what to say, how to act, and what to wear.

Are your people trained in public relations? Do your members understand the importance of public support to the fire department? Before allowing members to meet with the public there needs to be some type of training/education.

Making Your Financial Picture Clear To Potential Volunteers

An extension of clarity in words is the effort I urge you to make in telling people where your money comes from and where it goes. This clarity removes many people's reluctance to become involved and can establish trust ("They must not have anything to hide!") and encourage participation.

I would recommend a pie chart for depicting income and expenditures to allow people a graphic understanding of your figures. This makes your financial picture clear and easy to read.

Conclusion

Getting your message across is a vital part of marketing efforts. Many avenues are open to you, and you must carefully assess which will be most effective for each aspect of your work.

Carefully chosen volunteers with particular skills in communications, journalism, public speaking, broadcasting, etc. can be invaluable in your efforts.

In designing your messages keep the golden rule of communications in mind:

"Keep it simple; make it breathe!"

Notes ■

Training For Achievement

5

The Challenge — Decisions On The Fireground

On the fireground there are three basic types of decisions that must be made: Strategic, tactical, and task.

- **Strategic** decisions determine whether the operations will be of an offensive or defensive nature.

- **Tactical** decisions involve specific maneuvers and achievements of objectives.

- **Task** decisions center around the proper selection of techniques, methods, and skills.

All of these decisions are based on the capability of the fire response team to perform and properly execute evaluations so as to achieve results. It is absolutely imperative for the fireground officer to know the capability of their respective fire response teams. If these capabilities are not known, effective decision making cannot take place.

Specifically, how long does it take for your fire team to deliver a water flow of 400 gpm? 600 gpm? 800 gpm? How long does it take for your fire attack team to ladder a building with a 24' ladder and cut a 5'x5' ventilation hole in the roof? How long does it take for your crew to do breathing apparatus and advance a 2 1/2" hose line to the second floor of a building? These questions must be answered in order for the officer to effectively command the emergency scene and establish a plan of action to combat the fire.

The success of fire attack operations revolves around the ability of fire attack crews to execute certain basic skills and evaluations. These fundamentals must be executed with proficiency and consistency. The development of this high level of proficiency can only result from training.

The Importance Of Training

To be a good firefighter you must possess discipline, self-respect, pride in the department, a high sense of duty and obligation and the physical ability to perform. The

mission of the training division is to develop firefighters who will achieve results on the fireground/emergency scene by making the right decisions and executing skill and maneuvers proficiently and consistently.

Learning by experience, alone, is a slow process that can never lead to a broad subject of knowledge. While individual experiences may give an individual an adequate ability to perform, it certainly can never give an insight into the wide range of possibilities that are likely to be encountered during an emergency incident.

The function of a training program to be effective, it must develop the self-confidence in the individual to perform under stressful and hostile conditions. A training program must be systematic. It must be comprehensive. It must provide feedback to the trainee. It must be geared to achieve results.

Training Program Goals

There are three basic goals within a training program.

- **The first goal** is to develop a system, composed of people, apparatus, and equipment, that is capable of handling emergencies.

- **The second goal** of a training program is to meet community obligations. These are the expectations of a community expressed through utilization of existing services and the demand for new services within the community as it grows. The community has certain expectations of the fire department. These expectations must be met.

- **The third goal** of a training program is to satisfy the training needs of the individuals: the individual firefighters, technicians, and fire officers.

A training program will focus upon three areas:

- Skill development

- Skill maintenance

- Skill improvement

Skill development would be exemplified by basic training programs where a person comes in with little knowledge, goes through a program and develops skills and abilities. The second program would be one of maintenance in which knowledge retention and skill degradation are considered. If skills degrade, or if there is a problem in knowledge retention, then we're looking at a skill improvement program. That skill improvement program would be a remedial program in which deficiencies are noted and corrective actions are taken.

The Three "C's" Of Training

A person who has successfully completed any training phase or program should possess the three "C's": competency, confidence, and contributing.

- **Competency** in knowledge of technical information and subject to matter required of a fire protection professional. Competency in the execution and proficiency skills are required in the performance of their duties.

- **Confidence** in their own ability to perform under stress and adverse conditions. Confidence in the ability of the personnel they are working beside.

- **Contributing** to the success of fireground or emergency scene operations. Contributing to the success and growth of their department and the fire service.

The Challenge

An effective fire department training program is a challenge.

Training to develop a person's ability to perform a skill or maneuver is only half of training for achievement. Developing the self-confidence of the individual to perform those skills and maneuvers under extreme pressures and stress in hostile environments, such as a structural fire, at any time of day or night is the critical factor. In your training program: train to develop skill, train to develop self-confidence, train to achieve results of the fireground. Training is the key to the success of a fire department. You are the key to the success of your training program.

A training program will focus upon three areas:
- *Skill development*
- *Skill maintenance*
- *Skill improvement*

A person who has successfully completed any training phase or program should possess the three "C's": competency, confidence, and contributing.

Evaluating A Fire Department

6

Training Program — Eleven Key Terms

The Eleven Key Terms. What are the qualities of a successful training program? Eleven key items can be identified as significantly impacting a training program. Depending on their degree of involvement, these factors can become a strength or a deficiency.

The eleven key items impacting a training program:

1. Training division personnel

2. Training schedule

3. Training facilities

4. Training program objectives

5. Motivation for training from the administrative perspective

6. Educational methodology that is utilized for training

7. Company operations and performances

8. Reinforcement operations and performances

9. Personnel to be trained

10. Administrative priority regarding training

11. Peer group pressure and feelings of the fire department members

Key Item Number One

In examining this component for evaluation, the question must be asked: "Is there a training officer assigned?" And if so, does that person function as a manager? All too often training officers serve only the function of instructor. While in small departments this may be necessary, the major function of the training officer should be one of management, one of coordinating, one of providing continuity to the program.

This is the individual who will be involved with the establishment of objectives for specific training programs, and also providing feedback and evaluation to determine the effectiveness of the program itself.

Other items that come into play: Are there specified instructors for different subject matter? Examples would include: Driver training, pump operation, rope handling, self-contained breathing apparatus, hazardous materials, etc. A very important component of training division personnel is the first line supervisor.

The first line supervisor is accountable for the proficiency and abilities of his people. If this person is to be regarded as an instructor, however, difficulties can arise simply because of the connotation of the word "instructor." The first line supervisor, or the company officer, should be regarded as a "coach"; an individual who has a team which must be ready to play at any time. It is the responsibility of the coach to make sure that each person is maintaining a specified level of performance. It is the role of the coach to identify any deficiencies and to take corrective measures to bring them up to the acceptable performance level.

Key Item Number Two

The training schedule is a tool to help guide the training program along its route so that specified objectives can be obtained. Questions that must be asked:

- Is a training schedule posted?

- If so, how often?

- And more importantly, is the schedule adhered to?

Finally, are performance evaluations a part of that schedule? In many departments, training activities are on a very short term planning basis.

In some departments there are no formal schedules, yet informal schedules do exist. And still in other departments absolutely no schedule of any type exists.

The formal schedule should provide for a regular schedule of:

A. classroom sessions

B. outdoor training sessions

C. proficiency evaluations scheduled on a monthly, bi-monthly, quarterly basis, but no more than six months apart

These planned evaluations provide feedback, to the training officer and the company officer, on the ability of people to perform.

Key Item Number Three

While training facilities can become extremely expensive, the basic premise is that training facilities provide simulate conditions so that individuals and companies can practice and perform firefighting skills. Training facilities need not be elaborate, but they do need to be functional. A bare minimum should be a designated classroom and a paved or gravel area that is available for hose drill.

The importance of a designated classroom should not be over looked. A designated classroom provides a conductive environment to training and education. An apparatus room that is converted temporarily to a classroom does not possess desirable characteristics such as acoustics, lighting and temperature control. Plus the fact that it is a secondary use and not a primary use for training, which reflects upon training as an activity that is of secondary importance.

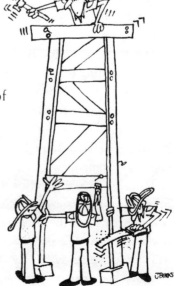

A paved area that is available for hose drills and operations is extremely important. This does not have to be on the fire department property, yet it must be available for use such as a church parking lot, shopping center that is not in use, an abandoned road, etc. Another important facility is a tower of some type. It could be concrete, cement block, or even a wooden pole tower. The objective is to provide an environment in which operations that take place aboveground can be simulated and practiced.

Other training facilities include a building in order to provide live fire training; a smoke maze in which self-contained breathing apparatus training can take place; and a flammable liquid pit for dealing with Class B fire.

When it comes to establishing training facilities, it is more important to come up with ingenuity, creativity and donations rather than dollars.

Key Item Number Four

Many times the objective is "to train." There are those programs in which the objective is to train "X" numbers of hours. Some objectives are traditional and unchanging. Some programs have objectives that are periodically revised.

Objectives must be achievable, and should be achieved on a regular basis. Of important note is that of documentation, to provide a tool for evaluating the program and to provide a tool of future goals and objectives.

Key Item Number Five

In your training program, is training passively supported by the administration as a minor activity? Is it supported simply to meet ISO recommendations? Possibly, it could be a traditional department activity which will insure adequate performances.

An effective and comprehensive program will have a strong administrative policy to train in order to develop individuals and companies to their potential, and to develop pride in their unit.

Key Item Number Six

Too often, the methodology which is used is lecture with minimum "hands on" training sessions. The key is to go from a methodology that is primarily instructor-oriented. Advance techniques should be employed, and innovative approaches, with the emphasis being learner-oriented (such as structured experiences), should be integrated into the program.

Learning is not passive activity. Learning does not take place by osmosis. Time must be allowed for an individual not only to learn, but to practice skills, especially in the manipulative areas.

Key Item Number Seven

The culmination of the training program manifests itself in the operations on the emergency scene. Fireground performance may be haphazard with operations that are not standardized and at showing the lack of formal evaluations. Sometimes there are standardized team evaluations, yet they are not adhered to. Do pre-fire plans exist but are not utilized? The training program should

strive for standardized team evaluations that are adhered to and periodically evaluated. Pre-plans should be quite evident: they should be systematic, and be revised periodically.

Key Item Number Eight

Often times, reinforcement is only of the negative type, in the forms of criticism, charges, and ridicule. If we are to expect quality performance, we must provide positive reinforcement. The training program should seek to recognize areas of deficiency, but also praise those areas in which performance is correct. The recognition should be active, not passive. Individuals should be recognized for their skill and companies should be recognized for their ability to perform.

Key Item Number Nine

To have only one specified level of trainee, such as firefighter, is quite inadequate. There needs to be different specified levels of trainees (i.e. recruit, firefighter, advanced firefighter, apparatus operator, and supervisor/fire officer). While each program does not have to be elaborate, it should provide basic orientation, understanding of their role within the organization, and training so that they can develop their skills and knowledge to be proficient in the role that they are performing.

Key Item Number Ten

Many times, administrative priority is only lip service. It can be evident, from the allocation of fiscal resources, what type of priority the administration places on training. If funding is inadequate or nonexistent, this can be a sign that there is a low administrative priority on this training.

Secondly, the amount of time allowed for training indicates administrative priority. Training takes time. If adequate time is not allowed, it is difficult for a training program to function effectively. If time is allocated and expected to be utilized in the proper manner for training, then it is evident that administration places a genuine priority on this activity. A training program must be sufficiently provided with time to achieve its objectives. A high administrative priority is evident by:

A. recognizing the factors of budget funding and time allocation

B. adequately providing them to support training activities

Key Item Number Eleven

If there is strong opposition to training by firefighters, there will probably be no input by the members of the department. In some cases, there may be mild opposition. Input in these cases would be rare or possibly nonexistent. If training is passively accepted, there will also be the correlation that occasional training may be initiated without directive.

Once training is accepted and expected, the input of the firefighters is an integral part of the program. Training will then be frequently initiated without directive. The feeling of department members and the peer group pressure can be utilized to support training activities, rather than to serve as a detriment and act against it.

Training Program Evaluation

Utilizing the following chart, evaluate your department's training program in each of the eleven key items. Rate each item from 1 to 9 according to which statement best describes the existing situation.

This evaluation can identify three points of reference for the development of a training program:

1. The department philosophy of the existing program.

2. The areas of strength relating to the existing program.

3. The areas where improvement is necessary.

Item	Rating
1. Training Division Personnel	_____
2. Training Schedule	_____
3. Training Facilities	_____
4. Training Goals/Objectives	_____
5. Motivation	_____
6. Methodology Utilized	_____
7. Company Operation/Performance	_____
8. Reinforcement	_____
9. Personnel To Be Trained	_____
10. Administrative Priority	_____
11. Peer Group Pressure	_____
Total	_____

Divide total points by 11. The resulting figure is a numeral rating from 1 – 9

Total Points	_____
Overall Rating	_____

Illustration 1 (see appendix)

RATING

	Key Items	1	3	5	7	9	Your Rating
1	Training Program Personnel	• no training officer • no instructors	• training officer position exists but no one assigned		• designated training officer as manager and designated instructors	• training officer is manager • specilized instructors • company officer instruct	
2	Training Schedule	• no schedule of training	• no formal schedule	• schedule every quarter but not fully adhered to	• formal schedule adhered to but no proficiency evaluations are scheduled	• formal schedule adhered to and proficiency evaluations are regularly scheduled	
3	Training Facilities	• no training facilities • no classroom	• apparatus room used as classroom	• designated classroom • paved/gravel area available for hose drill	• designated classroom • paved/gravel area • drill tower used regularly	• facilities used on a regular and systematic basis • integral part of training program	
4	Training Program Goals and Objectives	• none started	• only goal is "to train" • no specific objectives stated	• goal is to train 'X' amount of hours • objectives are unchanging	• objectives periodically revised but are not necessarily always achieved	• new objectives stated annually and objectives are achieved on a regular basis	
5	Motivation For Training	• none	• passively supported by administration as a minor activity	• supported in order to meet ISO recommendations	• a traditional department activity to ensure adequate performance	• strong admin. policy to train in order to develop individuals and companies to potential and pride	
6	Methodology Utilized for Training	• none	• primarily lecture with minimum hands on training sessions	• approx. 50% lecture and 50% hands on	• basically instructor oriented techniques with little innovation	• advanced and innovative techniques with the emphasis being learner oriented	
7	Company Operations and Perforance	• haphazard • no standardized and or formal evolutions	• no standardized team evolutions and no pre-planning	• standardized evolutions exist but not adhered to • pre-plans exist but not used	• standardized evolutions adhered to • pre-plans utilized and updated	• standardized evolutions adhered to and evaluated regularly • pre-pland are systematic	
8	Type of Reinforcement Utilized	• no positive recognition • negative reinforcement	• passive recognition • negative reinforcement	• passive recognition of individuals • some positive reinforcement	• active recognition of individuals with positive reinforcement	• active and regular recognition of program and individuals with positive reinforcement	
9	Personnel Specified to be Trained	• none	• only specified level is "firefighter"	• specified levels of firefighter and apparatus operator	• specified levels of firefighter recruit, firefighter, and aparatus operator	• various levels including recruit, firefighter, specialist/technicians and officer training	
10	Administrative Priority Regarding Training	• no time • no budget • no priority	• minimum time • no budget • no priority	• minimal time • minimal budget • low administrative priority	• time allocated • minimal budge • medium administrative priority	• time and budget allocated to meet training objectives • high administrative priority	
11	Peer Group Pressure Feelings of Fire Department Members	• strong opposition to training • no imput by members	• mild opposition • imput is rare or non-existent	• training accepted passively • training activity rarely initiated without directive	• training is accepted • training sometime initiated without directive	• training accepted and expected • input is integral part of program • training initiated frequently	
						Total	

A rating of 1 would indicate:

Training by apathy. In this type of condition, training is not taking place. Apathy prevails and consequently the operations within that department will be of a poor nature.

A rating of 3 would indicate:

Training by coincidence. In this style of training, abilities that are developed, knowledge that is gained, and proficiencies that are accomplished occur in a very haphazard nature. Training will be strictly by accident; it will not be planned, it will not be systematic. Training will result not from intent but merely by coincidence.

A rating of 5 would indicate:

Training by objection. Yes, training by objection – not objective. In a case like this, training is taking place. However, training is being passively accepted by the members of the department. It is being supported as something that must be done so "let's get it over with." Training by objection is a very minimal training program with minimal results.

A rating of 7 would indicate:

Training by tradition. This is a program that has a very solid foundation. It is a program that is well established and has been going on for quite some time. There is nothing at all wrong with training by tradition, except for the fact that it is not moving forward. And by not moving forward it is standing still. It is gradually losing ground over a period of time. However, training by tradition should be recognized as a program that is providing training of an acceptable nature.

A rating of 9 would indicate:

Training for achievement. This is at the other end of the spectrum from "Training by Apathy." This is a program that is very progressive, innovative, supported by the administration and accepted by members of the department. Because it is based on achievement, rather than merely fulfilling a function of training, it has greater motivation, greater interest, and the standard of performance will be much higher.

Notes ■

Training For Achievement

7

Critical Concepts For Success

Five concepts give continuity to "Training for Achievement." These concepts provide a perspective on the mechanics of training. The concepts take into consideration the human aspect of training fire personnel. The concepts are critical to the success of a training program.

Concept No. 1

Developmental training cycle

The Developmental Training Cycle consists of five segments:

1. Challenge
2. Work
3. Achievement
4. Reward
5. Rest

Weight training best typifies the Developmental Training Cycle. A person does not lift the same amount of weight the same number of times every day. If a person is to develop certain muscles, the weights are varied, the number of repetitions are varied, and there are periods of intense workout, and periods of rest.

A person must be challenged. The challenge must be realistic. Once a person has been challenged, or the expectations specified, a period of intense work is undertaken. During this period of work, a person begins to realize those expectations. At the end of the work period, an evaluation must take place. That evaluation measures the amount of achievement. Once this achievement has been recognized, a reward must be given. This recognition/reward reinforces the person's achievements.

Challenge

To effectively challenge a person is to identify a specific level of performance that is attainable by the individual. Yet, the challenge causes them to stretch in order to be successful.

A situation that would illustrate this point is the person who is physically sound but out of shape. At best, the person could run a mile in 12 minutes. The challenge would be to run that mile in 7 minutes. It is attainable by the individual. However, that individual must "stretch" themselves to be successful.

Work

In the work segment cycle, adequate time must be provided so that the required development can take place. The work must be scheduled on a regular basis. Progressive development must be programmed into the work process.

As in the illustration of our runner, he will schedule the number of weeks it will take to be able to achieve that 7 minute mile. The person will run and exercise on a regular and systematic basis. "Mileposts" may include being able to run a 10 minute mile, an 8 1/2 minute mile and finally a 7 minute mile.

During this work segment, the instructor performs the function of the "coach." The purpose of the coach is to assist the individual in achieving the specific level of performance. The coach does this by providing technical information, demonstrating proper technique and skills, encouraging the person, correcting deficiencies, adhering to the schedule and in general, supervising the person's activities.

Achievement

Achievement results when the person consistently performs the task to the standard set forth in the original challenge. This points up the importance of assuring that the challenge is attainable and able to be measured against a standard. In the case of our miler, it is a case of running the one mile distance against the stopwatch.

Reward

With achievement goes acknowledgement and reward. It is important to remember that the reward must be commensurate with the work required and the difficulty of the achievement.

Every achievement must be rewarded. A reward may be as simple as a pat on the back, an "official" letter of commendation, a certificate of accomplishment or a mention in the local newspaper. It may be as elaborate as a large trophy or gold medal. Department baseball caps, badges, shoulder patches and wall plaques are other examples of rewards.

Rest

The final segment of the cycle is rest. Yes, that's right; just plain rest. If the person has worked hard and achieved, they deserve a rest. The length of the rest segment, like the reward, should be commensurate with the work performed and the degree of accomplishment.

Mountain climbers illustrate this segment. Their objective is to climb to the peak. They climb to a certain point then rest. They climb further up the mountain to another point, then rest. They repeat this until they reach the peak and accomplish their objective.

In the training process, individuals will reach their own "peaks." It is important that rest periods are included, otherwise the risk of "burnout" severely increases.

Training cycles should last for 30 to 120 calendar days; 60 to 90 calendar days seem to work best.

An important aspect of the training cycle is that there is:

1. a definite beginning date

2. evaluation to measure the degree of achievement

3. a definite completion date

Concept No. 2

Performance standards

Performance standards, or objectives, are not new to the fire service. Performance standards are not, however, an end result or product. They are an indicator of proficiency. They are a "tool" for the training officer and the fireground officer.

A person wants to know what is expected of them and how well they are to perform a specific task or function. Performance standards are written statements that answer those questions.

Benefits of performance standards include:

- Standardize operations

- Provide uniformity

- Identify required skills

Performance standards apply to individual skills, single company evolutions and multi-company operations.

Performance standards consist of three components:

1. **Task** or evolution

2. **Conditions** under which the task or evolution is performed

3. **Criteria** that must be satisfied in order to successfully perform the task or evolution

When establishing performance standards, it must be emphasized that they are to be realistic in the time requirement. Performance standards are for fireground operations, not for "show" on the drill ground. The time requirement should reflect performance requirements under actual conditions.

The performance standard is **not** to be a speed contest.

To insure safety in the operation, certain basic rules should be observed during the practice and evaluation sessions:

1. Personnel should not leave the apparatus until it has come to a complete stop.

2. The apparatus should not take off until all necessary personnel are securely aboard the apparatus.

3. Unreasonable and abusive treatment of apparatus, hose, appliances and equipment should be avoided.

4. No running should be allowed. "Running" can be defined as a pace or speed greater than what it requires to travel 150 feet in less than 12 seconds.

Keep in mind that speed and proficiency result from accuracy, timing, coordination and many hours of practice.

Performance standards are for fireground operations, not for "show" on the drill ground. The time requirement should reflect performance requirements under actual conditions.

Keep in mind that speed and proficiency result from accuracy, timing, coordination and many hours of practice.

Concept No. 3

Performance evaluations

The purpose of evaluation is to measure progress. It is a means of providing feedback to the individual or group relative to their development. It is the means of correcting deficiencies before they become habits – bad habits. Performance evaluations are a tool. A tool to improve a person or group's proficiency and capability.

The attitude toward performance evaluation must be positive. All too often, evaluations, or testing, have been used as a form of ridicule, as a form of belittlement, or as a form of criticism. When evaluations are utilized in this manner, they have a very negative context and bring with them a negative learning environment.

Performance evaluations can function in three different ways:

1. **Inventory capabilities.** Performance evaluation can inventory performance capabilities of individuals and companies. The purpose is to establish a reference point of individual and team capabilities.

2. **Maintain performance level.** Performance evaluation sessions, conducted on a regular basis, can insure that established performance standards are maintained.

3. **Correct deficiencies/improve performance levels.** Performance evaluation sessions can identify deficiencies in skills. In this use of performance evaluation, it is the responsibility of the training officer or drill master to identify those skill deficiencies and prescribe selective training procedures to correct those deficiencies.

Performance evaluations are a tool. A tool to improve a person or group's proficiency and capability.

Concept No. 4

Challenge through enrichment

Promotional advancement is a form of challenge. However, not all persons want to advance upward nor will be able to advance upward. Yet persons need and want to be challenged. The enrichment of a specific position can be that challenge.

For example, take the position of driver/apparatus operator. The first step is being able to drive a standard pumper and operating the pump. The second step (enrichment) could be driving a larger apparatus (e.g. tender). The third step (enrichment) could be driving and operating an off-the-road piece of equipment. The fourth step (enrichment) could be driving and operating an aerial ladder apparatus. The fifth step (enrichment) could be master operator with the ability to perform annual pump test.

Each step, or enrichment, requires additional training and skill. Each step carries with it additional responsibility and status.

The position of Firefighter can have any number of steps or enrichment. Examples include, but are not limited to:

- Fire Recruit

- Probationary Firefighter

- Firefighter/Engineer

- Firefighter/EMS-Rescue Specialist

- Firefighter/Truckman

- Firefighter/Hazardous Materials Specialist

- Firefighter/Heavy Industrial Rescue Specialist

- Senior Firefighter

- Master Firefighter

Challenge through enrichment relates back to the developmental training cycle. The keys to successful utilization of both are scheduling, degree of challenge, and credibility.

The scheduling should allow the person to peak (achievement and reward) then take a breather (rest) but be presented with a new challenge before their interest drops off.

The degree of challenge presented must "push" an individual or group to achieve. Yet the challenge must be realistic and achievable. If the challenge is not realistic and unachievable, then frustration will result. That frustration will devastate your training program.

The steps (enrichment) must mean something when they are achieved. If a person can attain the step but not meet the performance standard, it "steals" that achievement from those that can meet the standard. To maintain the credibility of the program is integral to its success.

The degree of challenge presented must "push" an individual or group to achieve. Yet the challenge must be realistic and achievable.

Concept No. 5

Individual wants and needs

It is necessary to recognize the training wants of individuals and their needs. The involvement of people in planning gives them a vested interest in the program.

The role of the training officer, or manager of the program, becomes one of a negotiator. Negotiations should be a "win-win" arrangement, not a "win-lose" one. The negotiations must take place within the boundaries and guidelines outlined by the restrictions and expectations.

In developing the training program, the individual should be asked:
- "What training would you like to see this year?"

- "What areas have you noted that would be of value for us to train in?"

Also by doing this a person has "a piece of action" in the program. Once the wants have been identified, they should be negotiated against the needs of the organization. The person best suited to identify these needs would be the first-line supervisor. Once these needs and wants have been negotiated, a profile will emerge. This concept is further discussed in Chapter Eight.

Notes ▮

Training For Achievement

8

Program Management

This chapter provides the individual responsible for training with a blueprint for designing, developing and implementing an effective training program within their organization.

It is not intended as a text book for instructor development nor as a definitive work on educational methodology.

What has been put forth in this chapter could have aptly been called, "What every new training officer wants to know, but no one tells them."

Step One — Identify The Candidate To Be Trained

In order to establish training objectives, it is necessary to identify the candidate that will be trained. The identification of the candidate requires focusing on three basic items:

- The Purpose of the Candidate
- The Function of the Candidate
- The Assignment of the Candidate

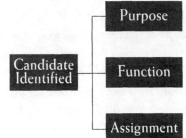

The Purpose Of The Candidate

The purpose of the candidate refers to their position in your organization's structure. Does the candidate function as a member of the labor force? Does the candidate function as a member of the management staff?

Two basic groups comprise the labor force. The first group is the new member or recruit. The second group is the skilled technician.

The recruit requires fundamental training in a specific skill area. Recruit training focuses on basic skill development.

Recruit training provides the basic knowledge and skill training that the candidate needs in order to be successful in performing their job. Further, it should develop the confidence of the individual to perform those skills proficiently under emergency conditions.

A basic premise in recruit training is that the candidate has the potential for success. This potential should be ascertained prior to the actual training. A valid selection process serves as a screening device which identifies those candidates with a probability of completing the required basic training successfully.

The skilled technician, on the other hand, possesses the basic skills. However, the proficiency of those skills must be maintained to a specific performance standard and enriched or supplemented so as to increase the technician's skill level and, ultimately, their value to the success of the organization.

A basic premise in training of experienced or skilled technicians is that the candidate be proficient in the basic skills required for the position.

Training for experienced individuals should expand their capabilities in a given

---

area. For this expansion to occur, there needs to be a solid base to build upon. Attempting to provide expansion training without the solid base of fundamentals would be equivalent to building a multi-story structure on a foundation of sand.

Training for experienced technicians must be provided to insure a consistent standard of performance. Training for experienced technicians need not be boring nor tedious. The use of simulations, practical exercises and periodic performance evaluations can be valuable in allowing individuals to demonstrate their proficiency and noting deficiencies which can be corrected before they manifest themselves in serious skill degradation.

Remember, success lies not in the complexity of the plan but in the consistent and successful execution of fundamentals. This was best epitomized by the great Green Bay Packers football team under the legendary Vince Lombardi. The Packers established a dynasty by executing the fundamental skills of running, blocking and tackling more consistently and successfully than any of their opponents. The success of your fire attack will not result from the complexity of the plan, but from the consistent and proficient execution of basic firefighting skills.

Just as there are basic skills for members of the labor force, there are basic skills for members of the management staff. The difference lies in the shift from manipulative skills (e.g. hose handling, ladder practices, etc.) to the conceptual skills (e.g. command, control, coordinate, etc.).

General principles still apply. Individuals new to the position must be provided with the fundamental skill training for that position and experienced individuals should be provided opportunities to expand their abilities.

Within the management staff for your organization are two broad categories: supervisors (e.g. company officers) and managers (e.g. staff officers).

A critical component in any organization and especially in a Fire Department is that of the first-line supervisor, known as the company officer. The importance of the company officer is equivalent to the knees of the career athlete. They are the pivotal point connecting the work force to the management staff. Poor company officers affect a Fire Department's performance in the same manner as bad knees affect an athlete's performance. All too often the "nozzelman syndrome" is applied to promotion of individuals to fire officers. The "nozzelman syndrome" translates to: "If a person is a great nozzelman, then they will make a good company officer based solely on their manipulative skill ability."

A company officer must have a sound background in fire combat. However, this should not be the sole determining criteria. Conceptual knowledge, human relationship skills and managerial abilities together with fire combat performance should be the parameters used to measure a candidate's ability to assume the role of company officer.

The Function Of The Candidate

Defining the function of the candidate further assists in establishing the identity of the candidate. The function addresses two issues:

1. What degree of supervision will be required by the candidate?

2. What level of group interaction will the candidate be operating at?

Will the candidate perform their assigned tasks under the direct supervision of an officer? Or will the candidate perform their assigned tasks under minimal supervision of an officer? How much judgement will the candidate be expected to display in the performance of their task as to selection of equipment and execution of various techniques?

The level of group interaction encompasses a wide spectrum. At one end of the spectrum is the ability of the candidate to perform as an individual. The other end of the spectrum is the ability of the individual to function effectively as a component or member of a team with individual actions and tasks being interdependent upon each other.

The levels of group interaction can be classed as:

Individual — The candidate to be trained functions as an individual performing a skill to achieve a specific objective or result. (e.g. A firefighter connecting hoselines to a hydrant and "charging" the lines.)

Unit — The candidate to be trained functions as a team composed of individuals operating as a single force performing a series of individual skills to accomplish a specific objective or result. (e.g. An engine company laying lines in from a hydrant, setting up a master stream appliance and establishing a large volume fire stream.)

Division — The candidate to be trained functions as a division, composed of units, operating as a single force performing a series of team skills to accomplish a specific objective or result. (e.g. An engine company at the water source supplying a second engine company operating attack hoselines while a truck company ventilates the structure, all units working to control the fire.)

Organizational — The candidate to be trained functions as an organization, composed of divisions, operating as a single force performing a series of team skills to accomplish a specific objective or result. (e.g. A large number of firefighters, apparatus and equipment working to control a large forest fire covering hundreds of acres.)

The Assignment Of The Candidate

The final step in identifying the candidate to be trained is establishing the assignment. What are they to accomplish? Most often, they are to accomplish:

- The suppression of fire

- The operation of emergency response apparatus

- Rescue

- Emergency medical aid

Consideration must be also given to:

- The prevention of fire

- Investigation of fire origin and cause

- Other activities that are necessary to your operation (support)

Identify the candidate to be trained

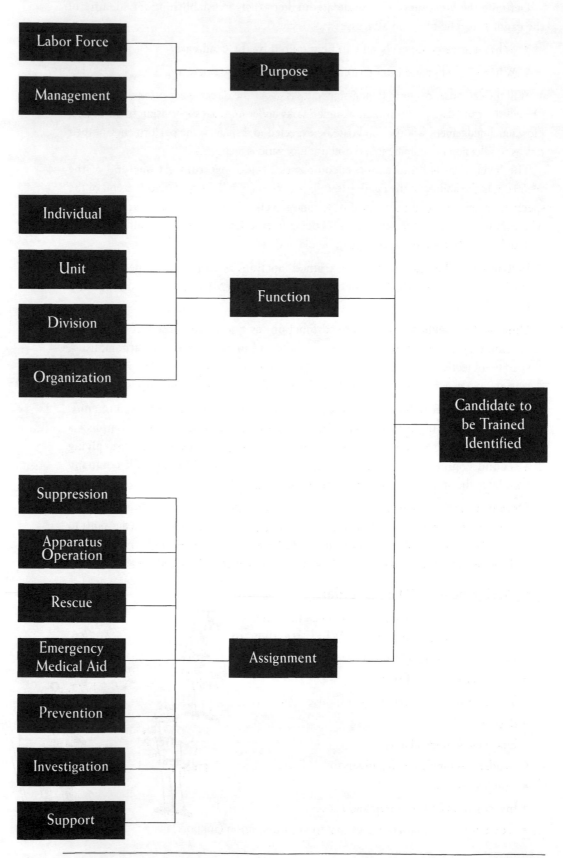

Step Two — Use Participatory Planning To Establish Training Goals

Participatory planning is a tool for making your training program successful.

In order for participatory planning to be effective, all three components must be in place, much like that of a triangle:

1. Training expectations

2. Service obligations

3. Program restrictions

■ Side One — Training Expectations

People have needs to be provided for and to be satisfied. Often this basic premise is not taken into consideration when developing a department training program.

Training expectations are the combination of the wants of the candidate and the needs of the candidate. Firefighters will identify their own training wants. Granted, it may not be realistic to accomplish all a candidate "wants." There may even be a few unrealistic "wants," but that must be evaluated by the individual in charge of the training program. The importance of having firefighters identify their training wants is that it provides for genuine input. This input allows the individuals who will be trained to buy in on the training program. They become personally interested in the success of the program.

The candidate has certain wants, as they relate to your training program. Based on the premise that people:

1. Want to accomplish

2. Want to achieve

3. Want to be proud of what they do

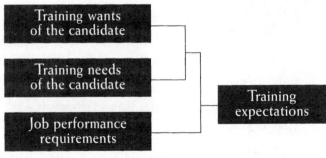

It is imperative that the "wants" be accurately identified. How do you do that? Easy! You ask the candidate.

Once the "wants" are identified, they should be compared to the "needs" of the candidate to be trained as identified by the candidate's immediate supervisor. A candidate can accurately identify their "wants" yet may not accurately identify all their "needs."

When identifying training needs, a cursory look should be given to job descriptions within the organization. Job descriptions may be as brief as a single paragraph or as complex as many pages.

Periodically, job descriptions should be evaluated against the actual performance required from a specific position. A major change in the job description of a "Firefighter" within the last ten years has been the upgrade of emergency medical services delivered by the fire department. An accurate analysis of job performance requirements may lead to a revision of the job description. The job description should be balanced against "training wants and needs."

To identify a candidate's wants and needs is not merely an exercise in administration; it allows the candidate to "buy-in" on the training that is to be presented. It recognizes the candidate and the individual responsible for the performance of the candidate – the supervisor. The ability of the candidate and supervisor to provide input to the design of the training program completes a very important process in the communication process – feedback. The end result is a desire to be involved in the training program, motivation to participate and consequently, to achieve.

The role of the training officer is to negotiate the "wants and the needs" against the job performance requirements. The result of this will be "training expectations."

Training Expectations   *Service Obligations*

Training Goals

Program Restrictions

■ Side Two — Service Obligations

The second side of the triangle is "service obligations." Your Fire Department and members are expected by different parties to be able to do certain things. The prime parties are the Fire Department administration, local governmental officials and the members of your community.

Administration has certain traditional obligations they become "endowed" with by virtue of position. Those obligations include statutory responsibilities which they are charged with. Further, obligations are put forth in general orders, policy statements, rules and regulations, etc. In some cases, Insurance Services Office or Survey and Rating Bureau Grading with specific class determination reinforce administration goals and obligations.

Administration goals and obligations need to be in line with community requirements. These requirements can be broadly categorized as:

- Immediate needs – less than 6 months

- Short range needs – less than 6 to 18 months

- Long range needs – greater than 18 months

And:

- What are the services currently being provided by your Fire Department?

- Are you providing only fire suppression?

- What level of emergency medical services is being provided?

- Specialized rescue services?

- Public education?

- Code enforcement?

- Hazardous materials special response?

The greater the range of services, the greater the responsibility to provide training to develop and maintain these skill levels.

If a "master plan for fire protection" has been completed for your community, a tremendous wealth of information is available at your fingertips. If one has not been done, involve other agencies and inventory your community needs and resources.

> *The role of the training officer is to negotiate the "wants and the needs" against the job performance requirements. The result of this will be "training expectations."*

Basic questions that should be answered in establishing a Fire Department's capability profile include:

- What services can your fire department provide?
- When can your fire department provide these services?
- What is the reflex or reaction time required to be operational from the time of discovery or notification of an emergency incident until personnel are on scene and operating?

In evaluating your department's actual capabilities, identify what tasks can be performed by:

- Individual firefighters
- Unit or single companies
- Division or several companies
- Organization or your total department

Tasks should be in a statement similar to a performance standard in which quality and quantity are identified. They must take into consideration special or unique needs of your community.

Once a department's capabilities are identified, it will be much easier to pinpoint your training objectives.

Department records, if reports are properly completed, can provide an accurate picture of services currently being provided. They can inform you as to:

- What services are being provided
- When they are being provided
- Where the services are being provided
- Who is receiving the services

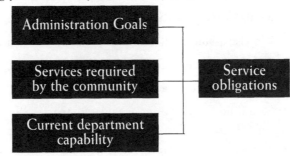

Further, they can provide a measure of effectiveness of services (e.g. **value at risk** minus **damage loss** equals **value saved or protected**).

Keep in mind, however, if reports are regarded as busy work and "filling in the blanks" is the sole purpose, then the accuracy of the reports and records are questionable, at best.

■ Side Three — Program Restrictions

The final side of the training triangle is often neglected. That side is "program restrictions." What do you regard as the biggest restriction placed on your training program?

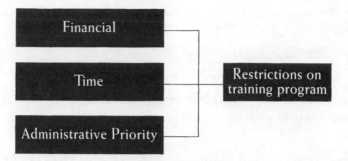

Financial Restrictions relate to:

- Payroll cost for fire personnel
- Contractual costs for outside training expertise
- Specialized training equipment
- Tuition, travel and subsistence for schools and seminars

Time restrictions relate to:

- Time to research subject matter
- Time to develop lesson plans
- Time to prepare training aids
- Time to deliver material to trainees
- Time to evaluate trainee's proficiency

Administrative priority restrictions relate to:

- Fiscal/Budget allocations in relation to other department programs
- Importance placed on training in relation to other function in the scheduling
- Adherence to performance standards

"Training Expectations" evaluated against "Service Obligations" then evaluated against "Program Restrictions" will result in a decision on training goals. A training goal is a general statement of something we want, some place we want to be or some future accomplishment which will further the mission of the organization.

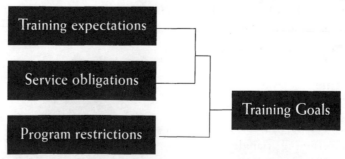

Step Three — Develop Training Program Objectives

A successful plan

To make training goals a reality requires a plan. In order for your plan to be successful, it must:

1. Have a direction or purpose

2. Have specific results to be achieved by specific dates

3. Be implemented

4. Be carried through to completion

Training goals

Training goals have three general directions or purpose:

1. To impart technical knowledge that the trainee needs in order to perform their job functions

2. To develop manipulative skills and proficiencies that the trainee needs in order to perform their job function

3. To encourage the proper attitude in the trainee toward their job function and tasks

In each of the three directions or purposes, it must be decided how simple or how complex they are to be. When the level of difficulty is determined, the training goal becomes better defined.

When writing training goals, the first question to be answered is: "**Why are we training?**" Are we training to:

1. Introduce and develop new skills and knowledge?

2. Review and maintain existing skills and knowledge?

Another question that must be asked is: "**Where do we start?**"

When an entry level is determined, a "starting point" is established. When the training objective is established, a "finish point" is established. With the start and finish known, the coverage of material is clearly outlined.

The entry level of the Candidate to be trained can be determined by:

• written evaluation

• task performance evaluation

• oral evaluation

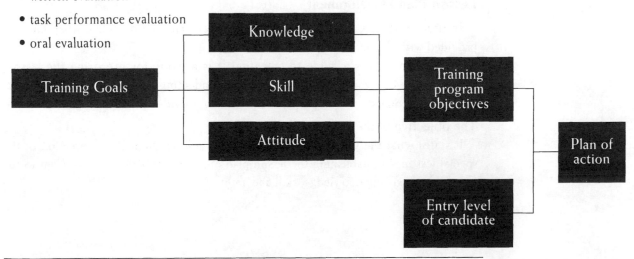

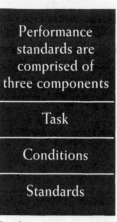

Performance standards are comprised of three components

Task

Conditions

Standards

Training objectives

A training objective is a specific statement that defines an end result to be accomplished by a specific date and, if necessary, at a specific cost.

Plan of action

In order to implement a training objective, a **Plan of Action** must be outlined. A plan specifies **Who** will do **What** at **Where** and **When** and **How** it will be done.

Necessary to the success of the plan is the provision for certain "controls." "Controls" include concise reports which explain how well the plan is being carried out. The reports tell you if the program is on schedule and if objectives are being achieved.

Performance standards

Performance standards are comprised of three components:

1. Task or action required

2. Conditions imposed while performing the task

3. Standards to measure the results against and determine if the results are successful

The performance standard (or often referred to as the Performance Objective) is critical to your training program because it provides a yardstick to measure progress in learning by the Candidate.

Further, a performance standard answers two important questions asked by the Candidate to be trained:

1. What am I expected to do?

2. How well am I expected to do it in order to be successful?

Common standards of acceptable performance:

- Speed – minimum time limits

- Accuracy – correct/without error

- Quality – proficiency/confidence

- Quantity – minimum amounts

Lesson Plan Development

In order for the Candidate to achieve the performance objective, they must be provided with the needed information.

The lesson plan is a guide to teaching. It is a guide for preparing the lesson, arranging the classroom or training area and conducting the delivery of the material in a logical sequence. A well prepared lesson plan provides the following information:

The objective of the lesson

It states what the Candidate will know or be able to do at the completion of the special lesson. In establishing the lesson plan objective, consideration is given to the Candidates' existing knowledge, skill and ability.

In order for the Candidate to achieve the performance objective, they must be provided with the needed information.

The organization of lesson material

- It identifies the major teaching points and presents them in a logical sequence or pattern.
- Subpoints, developing each of the major teaching points, are stated and control the scope of the lesson.
- Each of the teaching points (major and sub) bring the Candidate closer to being able to achieve the lesson plan objective.
- An explanation of how the lesson plan relates to job function and other lesson plans should be included to provide continuity in the training program.

The delivery of the lesson plan

The teaching method for effectively delivering the lesson material to the Candidate is identified. The teaching methods can include:

- Lecture
- Instructor-led discussion
- Demonstration
- Practical experiences
- Group interaction sessions (structured experience)
- Self-study (reading assignments)
- Individualized coaching

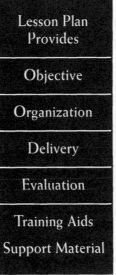

Lesson Plan Provides

Objective

Organization

Delivery

Evaluation

Training Aids

Support Material

The time required to deliver the lesson is specified along with a sequence schedule to move through the material at a certain pace and allow for coverage of material.

The classroom or training area arrangement is explained so as to facilitate the delivery of the lesson and allow the instructor to control and influence the learning climate. Included in the arrangements are:

- The number of students
- The arrangement of tables, chair, audio-visual equipment and props
- The need for assistant instructors and/or additional rooms if the class is to be broken up into smaller groups

Areas of the lesson plan which require special emphasis and how these areas will be stressed should be explained. Examples include lecture repetition, audio-visual aids, task repetition and distribution of printed material.

Of particular concern in delivering the lesson plan is the application of the material to actual job performances. Application to real situations not only clarifies the importance of the lesson, but also is a motivational tool to increase the Candidates' attention and ability to retain the information and techniques.

Evaluation of candidates' progress

Evaluation utilizing testing methods and devices provides feedback to the Candidate on the degree of success in accomplishing the lesson plan objectives.

The evaluation segment of the lesson plan should be identified and fully explained so that it can be properly utilized by the instructor. The three general categories of evaluation are:

- Written

- Oral

- Task performance

Training aids and support material

Training aids can be used to introduce, explain or summarize teaching points of the lesson plan. Training aids assist the instructor in emphasizing, illustrating or providing examples for clarifying ideas, equipment or techniques.

Support materials include student manuals, handout materials, controlled notes and other items necessary for effective instruction to take place. Other support materials which are advantageous include bibliography, reference materials and suggested readings to supplement the lesson plan. Training aids can be broadly categorized in the following classifications:

- Static — non-projectable

- Static — projectable

- Motion — projectable

- Motion — non-projectable

A chart is provided to assist the training officer/instructor in evaluating the need for specific training aids. To utilize this chart, it is only necessary to answer these four questions in order:

1. Is the display of motion required or necessary?

2. Is it necessary to utilize projection equipment?

3. Is it necessary to provide pre-recorded sound?

4. Is the display of color required or necessary?

Training aids which are static and non-projectable

Group 1
- Chalkboard
- Flip charts
- Flannel board, Magnetic board
- White board
- Printed material
- Black and white photographs

Group 2
- Chalkboard with color chalk
- Flip charts
- Flannel board, Magnetic board
- White board with color markers
- Printed material with color inks
- Colored photographs

Group 3
- Audio tape
- Printed material with audio tape

Group 4
- Audio tape with color printed material

Training aids which are static and projectable

Group 5
- Overhead transparencies

Group 6
- Overhead transparencies with color
- Color 35mm slides

Group 7
- Overhead transparencies with audio tape

Group 8
- Color 35mm slides with audio tape

Training aids which show motion and are projectable

Group 9
- Computer generated presentations, Black and white 8mm or 16mm film without sound
- Video tape

Group 10
- Computer generated presentations, Color 8mm or 16mm film without sound
- Video tape

Group 11
- Computer generated presentations, Black and white 8mm or 16mm film with sound
- Black and white video tape

Group 12
- Computer generated presentations, Multi-image slide projection with audio tape
- Color video tape
- Color 8mm or 16mm film with sound

Training aids which show motion and are non-projectable

Group 13
- Cutaway model of actual item
- Black and white graphic schematics with movable components

Group 14
- Cutaway model of actual item with color coding
- Color graphic schematic with movable components
- The actual item
- Simulator without audio tape

Group 15
- Color coded cutaway model of actual item with audio tape
- Color graphic schematic with movable components with audio tape
- The actual item with audio tape
- Simulator with audio tape

Develop Training Program Objectives

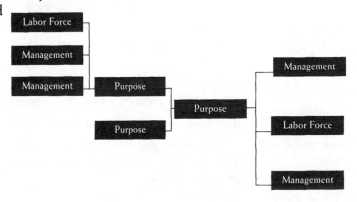

Step Four — Deliver the Training Material to the Candidate

Delivery of the training consists of six segments:

- Assemble
- Schedule
- Announce
- Present
- Evaluate
- Reward

Instructors, training aids, support materials and other resources need to be identified and assembled.

Times, dates, locations, equipment and instructors need to be coordinated and scheduled.

Scheduled training activities must be announced to all interested candidates. The announcement should serve to **Publicize** the scheduled training. It should **Advertise** not only the subject but the importance of the training. When announcing the training, attempt to trigger the candidates' interest and curiosity.

The presentation segment is the actual contact with the candidate. The key words to remember in presenting material are:

- Participation
- Positive reinforcement
- Progress

Learning is an active process. Both the instructor and the candidate must participate in the training activity. Every effort should be made by the instructor to increase the participation of the candidate.

With consideration given to the number of students, instructional methods that are learner oriented and maximize learner participation should be utilized. The learning climate must be conducive to achievement.

The instructor controls the learning climate. An important tool available to the instructor is positive reinforcement. Each correct response or achievement by the candidate should be reinforced. The reinforcement should be immediate and positive. Negative reinforcement will not eliminate undesirable actions. But positive reinforcement will let the candidate know they are achieving what is expected of them.

Training activities should move the candidate toward the achievement of the stated performance objective.

Regular evaluation measures the degree of achievement by the candidates and their progress in training process. Without a specified performance objective, it is difficult to accurately measure a candidates' progress.

The importance of evaluation cannot be overemphasized

Successful performance or achievement must be rewarded. Whatever the acknowledgement or recognition, it should be commensurate with the degree of achievement.

Step Five — Document the Training

After training is provided, document the training. Reports and records provide valuable information, if they are properly utilized. Reports and records can provide a barometer in gauging the progress of your training program.

If you currently have a record system that is working for you, continue using it. However, if you are not pleased with your record system, there are a wide variety of forms and software available. Whatever system you choose, keep two qualities in mind – simplicity and usefulness.

Each training session must be documented. If a training session is scheduled but not conducted, a training session report should still be filed with an explanation.

The illustration below is a sample training session report which can be put on a 5" x 7" card. The members in attendance can be continued on the back side of the card if necessary. The report can be filed by Company, Station, Battalion or whatever you desire.

| Document Training |
| Reports |
| Records |
| Results |
| Achievement |

Training Report   Date_____19____

Subject_____Hrs____

Instructor/Coach_____

Location_____

Company_____Officer_____

Members	Score	Company Members	Score
_____	_____	_____	_____
_____	_____	_____	_____
_____	_____	_____	_____
_____	_____	_____	_____

Evaluation used: Written   Oral   Task Performance
                        (Circle type used)

Comments:

Sample Training Report Blank

Training Report   Date *Feb 10* 19 *98*

Subject *Interior Fire Attack* Hrs *2*

Instructor/Coach *Assistant Chief Black*

Location *Station One Classroom*

Company *Eng. 521* Officer *Capt. Grey*

Members	Score	Company Members	Score
Grey	*Pass*	*White*	*Pass*
Quick	*Pass*		
Jones	*Pass*		
Brown	*Pass*		

Evaluation used: Written (Oral) Task Performance
                        (Circle type used)

Comments: *Classroom training session in preparation for live fire training drill in March.*

Sample Training Report Completed

Record of Training Blank

Company_____	Period From_____ to _____ 19____					
Record of Training **Company Members**	**Training Subjects Presented**					**Total Hours**

Record of Training Completed

Company _Eng. 521_ **(A)**   Period From **(B)** _Jan 1_   to _Feb 28_ 19 _98_

Record of Training Company Members **(C)**	Hose	Ladders	Pump Oper's.	Rescue Equip.	Interior Fire Attack	Total Hours
Captain Grey **(D)**	2 **(E)**	4	0	1	4	11
Engineer Quick	2	4	2	1	2	11
Firefighter Jones	2	6	4	1	2	15
Firefighter Brown	2	6	1	1	4	14
Firefighter White	2	6	0	1	4	13 **(G)**
Total Hours **(F)**	10	26	7	5	16	62 **(H)**

A   Specific groups who have received training

B   Specific training period

C   Training subjects during specific period

D   Individuals who have received training

E   Total training hours received by and individual in a specific subject

F   Total person hours received in a specific subject

G   Total training hours received by a specific individual

H   Total person hours of training provided

The individual reports can be compiled into records using a simple matrix. The following report provides the training record of "Eng. 521" for a two month period. This matrix can be used for:

- Weekly
- Monthly
- Quarterly (3 months)
- Semi-annual (6 months)
- Annual (12 months)
  or longer periods if necessary

This simple matrix provides you with statistical information on:

- Who has received training
- The amount of training received by an individual in a specific training subject
- The total training hours received by an individual
- The total training hours in a specific subject
- The total training in person-hours in a specific subject
- The total person-hours of training provided
- Training subjects provided during a specific period

Training Program Analysis

The reports and records provide statistical information that assists the training program manager in evaluating the progress and results of the training program.

When analyzing the training program, evaluate the "Three Rs:" – Reports, Records and Results/Achievements.

The analysis consists of four basic questions:

1. Are the end results satisfactory?

2. If the results are not satisfactory, which specific areas/objectives are not acceptable?

3. What caused these areas/objectives to be unacceptable?

4. What can be done to correct these problem areas/objectives?

The Annual Training Report

The documentation of training activities include the publishing of the Annual Training Report. The annual training report should include:

- Training goals
- Specific training objectives
- Degree of success achieved toward each objective
- Accomplishments of personnel
- Comments and recommendations

The reports and records provide statistical information that assists the training program manager in evaluating the progress and results of the training program.

When analyzing the training program, evaluate the "Three Rs:" – Reports, Records and Results/ Achievements.

A sample format of an annual report can be as follows:

- Title page
- Index page
- A cover letter addressed to the Chief of the Department or the governing authority
- A brief synopsis of training activity highlights during the year
- A review of the training goals and objectives for the year
- An examination of the achievement of the specified objectives
- Training program budget expenditures
- Actual training person-hour totals for the year
- Recommendations for future training
- Training program goals and objectives proposed for the upcoming year
- Training program budget for the upcoming year

The report should be word processed and placed in a folder with a cover. A sufficient number of copies should be provided to all interested and affected parties.

Notes ■

Training Division Personnel

9

Authority and Accountability– The Training Program Circuit

In the operation of a training division, there are four basic staff functions. In some departments, there are specific persons performing each of the four basic functions. In other departments, one person may perform all four functions. The importance is not who the specific individual is but, rather, the authority and accountability of the function in the "Training Program Circuit."

A circuit requires completion in order to be effective. An extremely important aspect of this circuit is that it provides feedback.

The first half of the training program circuit is the design, development and delivery of the training material. The second half of the circuit is the evaluation of students and analysis of training program results.

The staff functions involved in the circuit are:

- Administrator — Chief Training Officer

- Manager — Assistant Training Officer

- Technician — Training Specialist

- Instructor

Training Program Circuit — first half

During the first half of the circuit, the training division personnel function as follows:

Administrator — Chief Training Officer

- Planning and goal setting

- Evaluation of training needs

- Approves training program

- Initiates directional development circuit

- Prepares and oversees annual operating budget for training program

- Plans and develops new courses

- Spokesperson for training division

Manager

- Supervises staff personnel

- Designs program

- Develops training objectives

- Monitors training program progress

Training
Program
Circuit
(first half)

Direction

Design

Development

Delivery

- Schedules training
- Assigns instructors
- Prepares course outlines
- Analyzes need for instruction material

Technician

- Designs instructional materials
- Develops instructional materials
- Conducts research of subject matter for technical accuracy
- Conducts occupational analysis
- Conducts task analysis
- Writes lesson outlines
- Writes performance statements
- Writes examinations
- Develops audio visual materials

Instructor – Trainer/Coach

- Delivers training material to trainee
- Organizes and manages the learning environment
- Administers testing and conducts evaluation of trainee progress
- Supervises manipulative skill practice sessions
- Develops instructional materials

Training Program Circuit — first half

Training Program Circuit — second half

At this point in the circuit, the training material has been delivered to the trainee. This completes the first half of the circuit. During the second half of the circuit, the training division staff function as follows:

Instructor — Trainer/Coach

- Records test results
- Counsels trainees
- Completes training reports
- Acknowledges/rewards achievements of trainee

Technician — Training Specialist

- Analyzes testing devices
- Evaluates instructor techniques
- Evaluates teaching methods
- Revises instruction materials
- Revises audio visual materials
- Compiles training reports and records

Manager — Assistant Training Officer

- Evaluates program results
- Evaluates performance of training staff personnel
- Prepares and publishes periodic reports on the progress of training program
- Recommends future training programs
- Conducts quarterly performance evaluations

Administrator — Chief Training Officer

- Evaluates training program effectiveness
- Monitors division budget
- Evaluates performance of training division
- Conducts annual (or semi-annual) performance evaluations
- Prepares and publishes annual report on training activities
- Provides for storage and security of all departmental training records and reports

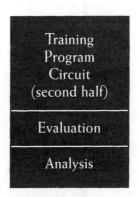

Training
Program
Circuit
(second half)

Evaluation

Analysis

Summary

At this point, the circuit is complete. The "administrator" specifies training direction and goals that Fire Department personnel will attain during the year. The "manager" establishes specific training objectives that must be accomplished if the goals are to be attained. Further, the manager supervises the training program and runs the day-to-day operation of the training division. The "technician" develops materials and provides specialized expertise. The "instructor" delivers the program and provides the direct link with the trainee. It is the instructor that transforms the training program from paper to performance.

You may wear one hat or all of the hats of the training division staff. What is important is the understanding of the different functions of training program management. In order for a program to accelerate, "shifting gears" is required. In this case, the gear shifting takes place when each member of the training staff fulfills their assigned function and responsibilities.

You may wear one hat or all of the hats of the training division staff. What is important is the understanding of the different functions of training program management.

Training Program Circuit — second half

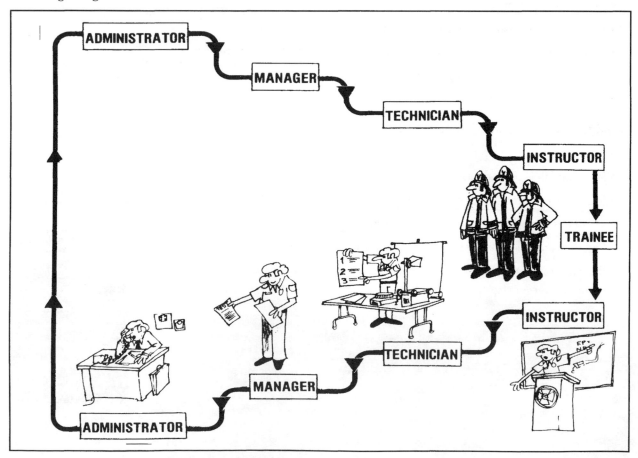

Model Programs

10

The Six Steps

The following programs and formats are provided as examples of programs which have been successfully implemented and achieved results in the training of Volunteer Firefighters. The programs included are:

1. Recruitment of Volunteer Firefighters

2. Selection of Volunteer Firefighters

3. Physical agility exam for Volunteer applicants

4. Training Volunteer Firefighters (40 hour recruit training program)

5. Probationary evaluation for Volunteer recruits

6. Integration of Volunteer Firefighter into an all-career (fully paid) Fire Department

Note: Consult with your legal counsel to ensure compliance with local state and federal laws prior to implementation.

"Success is a team effort"

Program One — Recruitment of Volunteer Firefighters

Goal

To recruit individuals as volunteer members of the fire department.

Objective

Recruit a minimum of (number) individuals who are interested in becoming a volunteer firefighter by (date, month, 19___).

Activities

1. Develop and present a slide program on the volunteer recruitment program to civic groups in local area.

2. Place ads in local newspapers to attract interested persons.

3. Gain exposure of the volunteer recruitment program through the local media.

Monitoring of program

Weekly progress report on the status of recruitment to be made by Assistant Chief/Training Officer to the Fire Chief.

Evaluation

Program shall be successful with a list of (number) individuals interested in becoming a Volunteer Firefighter by (date, month, 19___).

Program Two — Selection of Volunteer Firefighters

Goal

To select qualified individuals to participate in a Recruit Academy for volunteer firefighters.

Objective

Select a minimum of (number) individuals as qualified applicants by (date, month, 19___) to participate in a Recruit Academy for volunteer firefighters.

Activities

1. Preliminary screening of individual applicants according to following criteria:

   a) Age minimum of 18 years at time of appointment to Recruit Academy.

   b) Education requirement of High School diploma or equivalent.

   c) Be a citizen of the United States.

   d) Be of good moral character as verified by background check and personal references.

2. To successfully complete a physical agility examination comprised of jobs related to the performance of firefighting skills and impartially measure an applicant's ability to perform fire suppression duties.

3. To be interviewed by an oral interview board comprised of two members of the Fire Department and one member representing the local governing authority.

4. To undergo a physical examination and be certified in good physical health.

Monitoring of program

Status report of each applicant shall be forwarded to the Fire Chief by Assistant Chief upon completion of all four activities.

Evaluation

Program shall be successful with a list of at least (number) qualified applicants.

Program Three — Physical Agility Exam for Volunteer Applicants

Goal

To impartially measure applicant's ability to perform fire suppression duties.

Objective

Select applicants that can perform job-related tasks in the following activities:

a) Hoisting – Testing upper body strength

b) Coupling Hose – Testing manual dexterity

c) Ladder Climb – Testing upper leg strength

d) Hose Drag – Testing ability to pull to resistance

e) Balance Beam – Testing agility

f) Stairway Hose Advance – Testing stamina

Activities

Job related tasks to be performed by applicant in order to meet physical standards of Volunteer Program:

1. Hose Drag (dry) – Advance a 150' length of dry 2 1/2" hose a distance of 150 feet.

2. Ladder Lift and Carry – Lift a 24' extension ladder from the ground and carry a distance of 90 feet and place back on ground.

3. Hose Hoist – Applicant shall hoist a 50' section of 2 1/2" hose 30 feet upward and lower hose back down.

4. Balance Beam – Applicant shall walk a 4" wide beam for a length of 20 feet, turn in the opposite direction and return 20 feet to the starting point.

5. Stairway Advance – Applicant shall carry a folded and tied section of 1 1/2" hose up 4 stories while wearing a Self-Contained Breathing Apparatus tank on his or her back.

6. Tool Hoist – Applicant shall hoist a chainsaw, tied to a rope, a height of 4 stories, and lower back down to ground.

7. Advance of Charge Line – Applicant shall advance a charged 1 1/2" hose line from a marked point 35 feet in one direction, and 35 feet in a sideways direction.

8. Hose Coupling Drill – Applicant, when given the following hose and hose appliances, shall connect the various parts together to the best of their ability, after they have been shown a completed example of the assembled parts. Hose and appliances to be used will include: (1) 1 1/2" dry section of hose, (1) 2 1/2" dry section of hose, (1) 1 1/2" nozzle, (1) 2 1/2" to 1 1/2" gated wye.

Monitoring of program

Evaluators from the Fire Department will conduct the performance exam.

Evaluation

The applicant will be evaluated on a scale of either **Pass** or **Fail**. The program will be successful with a list of applicants that have successfully passed the physical task exam.

*Note: Care should be taken to insure that examinations are **Valid** as to the job related performance requirements within a specific Fire Department and **Reliable** to render consistent and objective test results.*

Program Four — Training Volunteer Firefighters

Goal

To train selected individuals in the performance of firefighting duties.

Objective

Selected individuals successfully completing 40 hours recruit academy by (date, month, 19_____).

Activities

1. Provide 40 hours of training in basic firefighting skills by attending and participating in:

   (a) Live fire training

   (b) Classroom lectures

   (c) Reading assignment

2. Instruction for recruit academy to be provided by:

   (a) On-duty fire department personnel

   (b) Training officer

   (c) State fire marshal's office (training division)

   (d) Related agencies

3. 40 hours of training to be scheduled on Saturdays (8 hr. day) of alternate weekends

Monitoring of program

Status reports by the assistant chief/training officer after 24 hours and 40 hours of instruction to be forwarded to the Fire Chief.

Evaluation

Objective is successful with (number) recruits successfully completing academy.
Recruit academy subject coverage:

- **Organization**: chain of command, size and scope of organization, standard operating procedures, work rules and regulations, terminology, radio code, alarms, communications, history, city areas.

- **Ladders**: safety, terminology, components, correct use, climbing procedures, ladder carries and raises.

- **Chemistry of fire**: use of water, exposure protection, types of fire, dangers of smoke and other toxic gases, heat transfer principles.

- **Hand tools**: identification of hand tools, location of hand tools, maintenance, ropes and knots, portable fire extinguishers.

- **Salvage**: theory, methods, tools, care, maintenance, and use of salvage tarps.

- **Breathing apparatus**: visual inspections, correct use, basic maintenance, nomenclature, proper wearing, introduction to smoke-filled environment.

- **Hose handling**: identification of hose, nomenclature, inspection and safety. Hose carries, rolls and folds. Handling charged lines.

- **Hose appliances**: types of nozzles, use of nozzles, master streams, hose appliances (wyes, siamese, reducers, male and female couplings). Correct use of a hydrant.

- **Ventilation**: theory, safety in ventilation procedures, technique, basic building construction, use of tools in ventilation.

Volunteer Training Program

	Session I	Session II	Session III	Session IV	Session V	Session VI
	1st Saturday	2nd Saturday	3rd Saturday	4th Saturday	Friday	5th Saturday
Morning 8 am - Noon	Protective Equipment and Safety 2 Hours	Chemistry of Fire 1 Hour	Chemistry of Fire 1 Hour	Chemistry of Fire 1 Hour		Training Exercise: Live-Fire Structural SESSION I
		Intro to Ladder 1 Hour	Hose 1 Hour	Ventilation Theory 1 Hour		
	Breathing Apparatus 2 Hours	Hand Tools 1 Hour	Hose Appliances 1 Hour			Training Exercise: Live-Fire Structural SESSION II
		Salvage 1 Hour	Hose Streams 1 Hour	Rescue Search 1 Hour		
			L u n c h			
Afternoon 1 pm - 5 pm	S.C.B.A. Drill 1 Hour	Ladder Drills 2 Hours	Hose Practice 1 Hour	Hose Ladder SCBA Rescue Search Drills 2 Hours		Training Exercise: Live-Fire Structural SESSION III
	Rope Class 1 Hour		Ladder Drills 1 Hour			
	Task Performance Practice and Evaluation 2 Hours	Rope and Hand Tools 1 Hour	Task Performance Practice and Evaluation 2 Hours	Hand Tools 1 Hour	Class Initial Fire Attack (7 pm) 2 1/2 Hours	Training Exercise: Live-Fire Structural SESSION IV
		Task Performance Practice 1 Hour		Task Performance Practice & Evaluation 1 Hour	Overhaul Tools 1/2 Hour	

Firefighter Recruit Academy

Recruit _____

**Tasks to be performed with proficiency
after 40 hours of basic training**

Tasks	Knows Task	Can Perform Task	Can Perform Proficiently
1. Identify and locate forcible entry tools on apparatus			
2. Ventilation using exhaust fan			
3. Ventilation breaking window or door glass			
4. Ventilation using water fog			
5. Tying hoisting knots: Bowline, clove, hitch, half-hitch, becket bend			
6. Tying rescue knots: Quick release bowline, butterfly knot			
7. Hoist tools: Axe, pike pole, fire extinguisher, roof ladder, saw, with tag line on each item			
8. Ladder carries: One person up to 24' extension ladder: Choice of (circle one)			
Arms length carry			
Shoulder carry			
Hi-shoulder carry			
Up to 35' extension ladder: (circle one)			
Arms length carry			
Shoulder carry			
Hi-shoulder carry			
9. Ladder raises: One person up to 24' extension ladder: Choice of (circle one)			
Hi-shoulder raise			
Shoulder raise			
Nordahl raise			
Ladder raises: Two person up to 35' extension ladder: Choice of (circle one)			
Beam raise			
Flat raise..			

Tasks	Knows Task	Can Perform Task	Can Perform Proficiently
10. Climb ladder to second floor			
11. Climb ladder while carrying tool			
12. Salvage cover throws			
13. Salvage cover folds			
14. Removal of water with squeege			
15. Rescue drag			
16. Searching for victims			
17. Use of breathing apparatus			
18. Inspection and care S.C.B.A			
19. Emergency breathing (with S.C.B.A.)			
20. Attack class "A" fires			
21. Identification of hose			
22. Identification of hose appliances			
23. Advancing dry hoseline into a structure			
24. Advancing dry hoseline up a stairway			
25. Hose roll – One person			
Donut roll (2 1/2" – 3")			
Double donut roll (1 1/2")			
Straight roll (1 1/2" – 2 1/2" – 3")			
26. Advancing hoseline (evolutions)			
Shoulder carry (dry)			
Wet (charged lines) 1 1/2" and 2 1/2"			

Program Five — Probationary Evaluation for Volunteer Recruits

Goal

To evaluate performance of volunteer recruit firefighter.

Objective

Qualified recruits shall successfully complete probationary period of 6 months by (date, month, 19——).

Activities

1. Recruits attend weekly drill sessions (2 hrs. per week in evening).

2. Recruits successfully complete written and task performance examination.

3. Recruits respond to 50% of alarms sounded to alert volunteer firefighters.

Monitoring of program

Monthly status reports on progress of program and individual recruits to be completed by Assistant Chief/Training Officer and forwarded to the Fire Chief.

Evaluation

Program is successful with the completion of probationary period by (number) recruits.

Program Six — Integration of Volunteer Firefighters Into an All-career (fully paid) Fire Department

Goal

To integrate selected and qualified individuals into the fire suppression division of the fire department as active and participating members of the organization.

Objective

Integrate volunteer recruits into the fire department as full fledged volunteer firefighters by (date, month, 19——).

Activities

1. Appoint staff officer as volunteer coordinator.

2. Establish a board of review.

3. Organize volunteer firefighters association.

4. Establish, define and publish departmental operating procedures relating to the following areas of concern:

   (a) Organization and chain of command

   (b) Response of volunteers to structural fires and staging area

   (c) Discipline code and grievance procedure

   (d) Station house rules, operation and activities

   (e) Individual firefighting performance skills

   (f) Engine company performance skills

   (g) Department terminology

5. Issue protective equipment, uniforms and alerting device to qualified and selected applicants.

6. Issue "letter of congratulation" to selected individuals.

7. Issue a shoulder patch and certificate to graduates of recruit academy.

8. Issue a badge and certificate to volunteers successfully completing probationary period.

Monitoring of program

Monthly status reports of volunteer recruits will be made by volunteer coordinator and sent to the Fire Chief.

Evaluation

Program will be termed successful with (number) active volunteers performing fire suppression duties by (date, month, 19——).

Motivating Volunteer Firefighters

11

It Starts With Good People

A fire department's ability to maintain volunteer firefighters is directly related to their ability to manage people. To maintain volunteers effectively, an organization must continuously work towards providing an environment which, by its very nature, tends to attract and keep qualified people. A program which is going to attract and keep good people must meet four (4) basic criteria:

1. The program must meet individual needs

2. The program must provide its membership with reward and recognition

3. The program must provide adequate supervision and leadership

4. The program must challenge members

As mentioned earlier, volunteers will come to your organization in search of means to fulfill their basic needs. These needs include:

- a sense of belonging
- achievement
- economic security
- freedom of fear
- love and affection
- self respect
- understanding
- challenge
- recognition
- reward
- the need to have fun and enjoy life

The basic need to belong and be affiliated with something that is valued is a very important need which the fire service can easily meet. By the very nature of the service we provide and the general attitude of most communities toward their fire departments, we can not only attract persons looking to fulfill their basic need, but we can do it much better than most other organizations. People like to be associated with a "winner" and if the fire department is doing its job, it will be a winner in the eyes of the public. Achievement can be experienced by volunteer members by providing opportunities for individuals to assist in developing, implementing and completing various tasks and projects. These opportunities come in many forms. They range from specialized training programs to organizing social functions. The challenge is not so much finding ways of providing members a chance to participate and achieve, as it is fitting the challenge to the individual so they will feel a sense of "achievement." Love and affection can be developed within an organization through years of working, laughing, crying and helping each other. One doesn't realize how much love and affection there is between department members, many times, until a member is removed. So was the case with volunteer captain Wes Claflin and the members of the Jackson Fire District No. 3 (to whom this book is dedicated). Self respect is developed through growth as an individual. The fire department provides considerable growth opportunities.

By the very nature of the service we provide and the general attitude of most communities toward their fire departments, we can not only attract persons looking to fulfill their basic need, but we can do it much better than most other organizations.

Important motivational areas include:

- treat people as individuals
- praise sincerely
- promote participation
- make work interesting
- promote cooperation and teamwork
- provide growth opportunities

Challenge, recognition and reward are the bread and butter of any program. If you want to stimulate a group, "challenge" them. If you want to continually stimulate them, provide recognition and reward for performance.

An annual awards banquet is probably the ideal activity designed to provide members with deserved recognition and reward. Many successful organizations feel that the annual awards banquet is the most important non-emergency activity in which a department can participate.

Last but not least, people have a basic need to have fun and enjoy life. The progressive volunteer organization provides opportunities for its membership to meet this need. Games, social activities, and friendly competition help meet this need.

A volunteer program which addresses these needs and provides for the achievement of these needs is going to experience success in maintaining volunteer firefighters. Many of the above-mentioned needs are accomplished in a basic program, including in-service training and responding to emergency situations. Other needs must be met by providing additional activities for department members.

It is important to remember that volunteer firefighters are individuals. They are individuals inasmuch as they have specific needs many times common only to themselves. Variety is an important factor in developing a comprehensive volunteer program. The organization must provide enough diversification of activities and events so as to get everyone involved and provide an important role for all participants.

Not only is it important to understand and meet the needs of the individual, it is also necessary to understand group needs.

Characteristics of a motivator

There is not a formula for motivating people, but it is important to remember the following: communicate and praise; consult with others; encourage participation; counsel regularly; and think in terms of a team.

Some characteristics of an effective motivator are:

- establishes goals and objectives that are worthwhile, challenging, attainable
- makes decisions after getting input
- seeks and gives feedback
- resolves conflict
- communicates regularly
- listens
- is open and sincere

Even once a high level of enthusiasm and participation is reached, it takes good managers and leaders to continue influencing volunteer personnel in a manner which is conducive to achieving goals and objectives with optimal efficiency and effectiveness.

To understand motivation and its role on performance is a key to organizational success.

- controls temper

- open-minded

- makes work interesting

- isn't afraid

- doesn't compete for credit

Some factors that cause us to be the way we are include:

1. Physical characteristics

   a. age

   b. sex

   c. weight

   d. height

   e. race

   f. physique

2. Childhood

   a. environmental conditions

   b. family unit

   c. care and treatment

There have been many studies conducted relative to concerns and priorities of employees and supervisors. One thing is apparent – the lists are different for each group as illustrated in the following chart:

Volunteer vs. Supervisor Job Priorities

Employee Ranking	Item	Supervisor Ranking
1	appreciation of a job well done	8
2	feeling of being "in the know"	10
3	sympathetic help on personal problems	9
4	job security	2
5	good wages	1
6	interesting work	5
7	promotion and growth	3
8	personal loyalty	6
9	good working conditions	4
10	tactful discipline	7

Motivating the individual volunteer

The need or desire some individuals have to be affiliated with something that is valued can be met by the fire service. The valuable service provided by fire departments can attract volunteers. Quality fire departments are viewed as winners in the eyes of the public.

In order to motivate people to volunteer, we must understand volunteers. Reasons given for volunteering include:

There is not a formula for motivating people, but is is important to remember to:

Communicate and praise

Consult with others

Encourage participation

Counsel regularly regarding teamwork and opportunity

- I wanted to help
- gain experience
- thought I would enjoy it
- popular activity in my town
- friends talked me into it
- the thrill of it
- I wanted to help people
- like working under pressure
- my father was a firefighter
- community service
- camaraderie
- training opportunities
- help keep taxes down
- social aspects

Volunteerism has changed over the past several decades. A favorite example of this change is the technique used by the United States Army to recruit volunteers for the military in the late 1940's and 1950's. "Uncle Sam" was prominently displayed on billboards, posters, and pamphlets. Today an advertising campaign like Uncle Sam would be unappealing or even humorous to young people.

The point is, our methods of attracting and motivating volunteer firefighters must change as people change. Recruiting programs will be effective only if we understand people, their needs, desires, and what turns them on and off.

Methods of showing appreciation

Recognition is not just a way of saying thank you. It is a valid response for volunteer services and provides a reason for becoming a volunteer. Recognition should happen automatically, particularly if the atmosphere is friendly. A simple smile is appreciated by any volunteer.

People are motivated by a variety of things. Volunteer program supervisors will learn what is important to their volunteers if communication is open and listening is practiced. The following outline, taken from Judy Raumer's book entitled "Helping People," suggests how recognition and motivation can be linked.

Achievement

- Letter of recommendation listing the individual's accomplishments
- Training opportunities that encourage personal growth
- Letters of appreciation to the volunteer's employer

Challenge

- Involve individuals in problem solving
- Use in emergencies
- Provide interesting job opportunities

Creativity

- Invite ideas regarding their own job placement
- Have a suggestion box
- Listen to volunteer ideas

Independence

- Effective orientation and training
- Opportunity to work without close supervision
- Let volunteers know they are trusted

Interest

- Match volunteers and jobs
- Sponsor attendance at community based workshops
- Avoid routine assignments

Leadership

- Ask to help train other volunteers
- Put in charge of project
- Invite to participate in planning and evaluation

Recognition

- Have a *Volunteer of the Month*
- Submit a name(s) for community level volunteer award
- Give certificate of appreciation at annual ceremony

Security

- Plan carefully
- Have task materials ready
- Say we missed you when absent
- Have name tags

Self Expression

- Invite volunteer to fill out interest and talent inventory
- Use talents and interests
- Environment receptive to volunteers taking initiative

Service

- Post letters of appreciation
- Record and publish ways and number of people helped by the volunteer program
- Training so meaningful contribution is made

Socialization

- Have get acquainted staff and volunteer gathering
- Provide lounge and coffee
- Holiday party for volunteers

Variety

- Supervisors should be creative and responsive
- Make job changes optional
- Check to see how things are going

Basic Maintenance Criteria
Meet individual needs
Provide members with reward and recognition
Challenge the membership
Provide adequate supervision and leadership

Motivation theories

There are several theories of motivation, each containing a list of individual needs that may be used for motivational purposes. Listed below are the basic human needs as stated in three widely accepted motivational theories.

Maslow's Hierarchy of Needs

- physical needs
- safety and security needs
- social needs
- esteem needs
- self-realization

Murray's List of Needs

- abasement–to admit inferiority
- achievement–to accomplish
- affiliation–to please and win affection from a valued object
- aggression–desire to overcome or fight
- autonomy–to be free
- counteraction–to overcome weakness
- defendance–to defend against assault
- deference–to conform or admire superiors

- dominance–to influence or control others

- exhibition–to be seen and heard

- harm avoidance–to avoid harm

- order–to put things in order

- play– to have fun

- rejection–to separate from negatives

- succorance–to have needs gratified by the sympathy of others

- understanding–to be understood

Rath and Burrell's List of Needs

- belonging

- achievement

- economic security

- freedom of fear

- love and affection

In her book entitled "The Effective Management of Volunteer Programs," Marlene Wilson discusses people motivated by achievement, power, and affiliation.

These personalities are briefly discussed below.

Achievement Motivated Person

Goal: Success in a situation which requires excellent or improved performance.
Characteristics:

- Concern with excellence and wanting to do personal best. Sets moderate goals and takes calculated risks.

- Likes to take personal responsibility for finding solutions to problems.

- Has desire to achieve unique accomplishments.

- Restless and innovative, takes pleasure in striving.

- Wants concrete feedback.

Spends Time Thinking About:

- Doing job better.

- Accomplishing something unusual or important.

- Advancing their career.

- Goals and how they can attain them. Obstacles and how they can overcome them. To illustrate, Sir Isaac Newton was once asked how he ever discovered gravity. He matter-of-factly replied, "By thinking about it all the time!"

Power Motivated Person

Goal: Having impact or influence on others.

Characteristics:

- Concern for reputation or position and what people think of their power and influence.
- Gives advice, sometimes unsolicited.
- Wants their ideas to predominate. Like Archie Bunker when he said, "It's times like this, Edith, where the only thing holding a marriage together is the husband being big enough to step back and see where his wife is wrong."
- Strong feelings about status and prestige.
- Strong need to influence others to change other people's behavior.
- Often verbally fluent, sometimes argumentative.
- Seen by others as forceful, outspoken, and even hard-headed.

Spends Time Thinking About:

- Influence and control they have over others.
- How they can use this influence to win arguments, change people, gain status and authority. McClelland, however, in an article entitled, "Two Faces of Power," published in the *Journal of International Affairs* in 1970, points out some mistaken notions we have in this country concerning the need for power. He stated we have almost totally overlooked the fact that power has two "faces"– one negative and one positive.

We tend to assume any leader with power must have dominated the group and attained their power at the expense of others. This is sometimes true, but not always. McClelland identifies the negative type of power as "personalized power." He calls the positive "socialized power." He characterizes each as follows:

Personalized Power (Negative):

- I Win–You Lose
- Law of the jungle
- Prestige supplies (e.g., biggest desk, nicest office, newest car)
- Makes group dependent and submissive
- Exerts personal dominance
- Tends to treat people like pawns, not origins

Socialized Power (Positive):

- I Win — You Win
- Exercises power for benefit of others to attain group goals
- Charismatically inspires others to action
- Creates confidence in others; helps them achieve group goals
- Makes people feel like origins, not pawns

McClelland cautions that if our society insists on overlooking or disregarding the positive use of power, we will continue to see our young people shun public office and positions of leadership. It would seem evident that we must start to reinforce the positive face of power, or in the future we will have even more difficulty getting good volunteers to run for school boards, city councils, university regents, city and county planning boards. Like so many other things, power is not bad in and of itself, it is the misuse of it that is.

Affiliation Motivated Person

Goal: Being with someone else and enjoying mutual friendship.

Characteristics:

- Concerned with being liked and accepted in inter-personal relationships.
- Needs warm, friendly relationships and interaction.
- Concerned about being separated from other people, definitely not the loner.

Spends Time Thinking About:

- Wanting to be liked and how to achieve this.
- Consoling or helping people.
- Warm and friendly relationships.
- The feelings of others and themselves.

Administrative implications

The administrative implications of McClelland's and Atkinson's motivation theory we have just described are quite dramatic:

1. Managers can select people whose motivational drive fits the job to be done.
2. Or vice-versa, we can fit the job to the motivational needs of the worker.
3. Managers can do things to work situation or organization which will help get the job done.
4. Managers can change themselves relative to how they lead others.

General tips for motivating volunteers

- Visit at station
- Call by name
- Give meaningful work
- Ask their opinion
- Put them on display
- Express appreciation regularly
- Stress their importance
- Send "Thank You" notes
- Be available when needed
- Dress them up
- Provide good leadership
- Be optimistic
- Have high expectations and goals
- Post results (attendance/performance)

Your catalog of motivational strengths

- Leadership
- Ability to follow
- Production of new ideas
- Ability to create
- Writing skills
- Speaking skills
- Quick learner
- Ability to convince
- Ability to organize
- Ability to make friends
- Ability to anticipate needs (Reprinted from *Nation's Business*, August, 1961.)

A few thoughts on self-motivation

- People utilize 2% to 5% of their ability.
- When your image improves, your performance improves.
- There is little you can learn from doing nothing.
- Your attitude toward someone affects their results.
- Criticize the performance not the performer.
- Be careful of "tradition."
- We treat people the way we see them.
- Go as far as you can. When you get there, you'll be able to go farther.
- Obstacles can be stepping or stopping stones.
- You're a lousy somebody else, but you're the best "you" in the world.
- Nobody can make you feel inferior without your permission.
- Success is easy once you believe in yourself.
- You get the best from others by giving the best of yourself.
- If you give a man a fish, you feed him for a day. Teach him to fish, you feed him for a lifetime.

Characteristics of good goals

- They must be big to us.
- They must be long range.
- They must be daily.
- They must be specific.

Volunteer burn out

We have those volunteer organizations which respond to a tremendous amount of calls and have an exposure rate that is adverse to long term employment of a volunteer due to too much activity. A volunteer who is called upon more than expected will ultimately be faced with a problem just as serious as the volunteer who is not getting enough activity. This situation is commonly referred to as volunteer burn out. Volunteer burn out occurs when a volunteer is involved to the point where it is no longer enjoyable. A volunteer who overexposes themselves also risks causing problems at home or with their chosen job or profession. In these situations, we must assist volunteers in pacing themselves. Encourage them to become involved only at a level which is comfortable. One way to avoid this situation is to balance the workload in the department through specific job assignments. Assigning volunteers to specific companies who are programmed and toned out is also an effective way of pacing a volunteer.

Volunteer burn out occurs when a volunteer is involved to the point where it is no longer enjoyable.

Counseling becomes an important part of dealing with volunteer members in your organization. Sitting down with the volunteer individually on a regular basis to discuss performance and general outlook is an effective means of avoiding problems.

Motivational tools for daily use

Delegation. The act of giving responsibility, with the freedom to act in an atmosphere of approval.

Access to information. Receiving or having available whatever information is needed for the successful achievement of one's responsibility.

Freedom to act. Acting on one's own initiative without interference from supervision in the form of "overcontrol".

Atmosphere of approval. The climate resulting from supervisory behavior which reflects confidence in the individual's competence and integrity.

Involvement. The participation and identification resulting from having an important role in the achievement of group goals.

Goal setting. Exercising initiative and judgment in establishing job goals for self or group.

Planning. The development of strategies and tactics for achieving job goals of self or the group.

Problem solving. The process of applying one's initiative, knowledge, and skills in the achievement of job objectives.

Counseling becomes an important part of dealing with volunteer members in your organization. Sitting down with the volunteer individually on a regular basis to discuss performance and general outlook is an effective means of avoiding problems.

Work simplification. A formalized training program which enables personnel to exercise their initiative and creativity for increasing job effectiveness.

Performance appraisal. A day-to-day developmental process which involves people in goal setting and work planning. This should be formalized as a semi-annual program for involving employees in the review of their job, appraisal of their performance, and planning goals.

Discretionary award. A special award given for meritorious performance or because of some unique contribution to the success of the company.

Inventions. Creating new procedures and processes for the achievement of individual and company objectives.

Transfers and rotations. Lateral transfers or job rotations to challenging assignments of equal responsibility.

Education. The development of employee knowledge, skills, and attitudes.

Memberships. Identification with interest groups through association with professionals, participation in meetings, and access to professional publications.

Utilized aptitudes. Responsibilities which utilize the individual's talents and satisfy their desire and capacity to grow.

Work itself. An assignment which affords gratification to the individual through the nature of work performed.

What firefighters want in an officer

1. To know the officer really wants them on the crew.

2. Helps subordinates learn their jobs.

3. Explains what they want.

4. Tells subordinates how they are doing.

5. Encourages growth.

6. Takes a personal interest.

7. Listens to other's ideas.

8. Honest with people.

9. Has faith in their subordinates.

What officers want in a firefighter

1. Knows and likes their job and tries to improve.

2. Keeps physically and mentally fit and alert.

3. Strives to get ahead.

4. Has pride in what they do.

5. Keeps the team spirit.

6. Seeks to learn.

7. Faces responsibility squarely.

8. Puts themselves in officer's place.

Volunteer viewpoint

If you want my loyalty, interests, and best efforts, remember that...

1. I need a "Sense of belonging," a feeling that I am honestly needed for my total self, not just for my hands, nor because I take orders well.

2. I need to have a sense of sharing in planning our objectives. My need will be satisfied only when I feel that my ideas have had a fair hearing.

3. I need to feel that the goals and objectives arrived at are within reach and that they make sense to me.

4. I need to feel that what I'm doing has real purpose or contributes to human welfare – that its value extends even beyond my personal gain.

5. I need to share in making the rules by which, together, we shall live and work toward our goals.

6. I need to know in some clear detail just what is expected of me – not only my detailed task, but where I have opportunity to make personal and final decisions.

7. I need to have some responsibilities that challenge, that are within range of my abilities and interest, and that contribute toward reaching my assigned goal.

8. I need to see that progress is being made toward the goals we have set.

9. I need to be kept informed. What I'm not up on, I may be down on. Keeping me informed is one way to give me status as an individual.

10. I need to have confidence in my superiors – confidence based upon assurance of consistent, fair treatment, or recognition when it is due, and trust that loyalty will bring increased security. In brief, it really doesn't matter how much sense my part in this organization makes to you – I must feel that it all makes sense to me. I would add, hopefully, that it all makes sense to everyone involved – the client, staff, volunteer, and you. (Reprinted from *The Effective Management of Volunteer Programs*, by Marlene Wilson, 1976).

Rights of the volunteer

A volunteer has the following rights:

- To be treated as a co-worker.
- To be given suitable assignments.
- To know as much as possible about the organization.
- To receive training.
- To receive continuing education.
- To receive timely feedback.
- To be given sound guidance and direction.
- To be given opportunities for advancement and varied experiences.
- To be heard.
- To be recognized.
- To be appreciated for their contributions.

Responsibilities of the volunteer

A volunteer has the following responsibilities:

- To be sincere in the offer of service and believe in the value of the job to be done.
- Loyalty to the community service they work with.
- To maintain the dignity and integrity of the community service with the public.
- To carry out duties promptly and reliably.
- To accept the guidance and decisions of the coordinator of volunteers.
- To be willing to learn and participate in orientation, training programs, meetings, and to continue to learn on the job.
- To understand the function of the paid staff, maintain a smooth working relationship with them, and stay within the bounds of volunteer responsibility and authority.

Conditions existing in an effective team

There are five basic conditions which exist in an effective team:

1. **Mutual trust.** Mutual trust takes a long time to build and can be destroyed quickly. It is established in a team when every member feels free to express their opinion, say how they feel about issues and ask questions which may display their ignorance, and disagree with any position, without concern for retaliation, ridicule or negative consequences. Trust between members is critical to the success of a volunteer program.

2. **Mutual support.** Mutual support results from group members having genuine concern for each other's job welfare, growth and personal success. If mutual support is established in a team, a member need not waste time and energy protecting themselves or their function from anyone else. All will give and receive help from anyone else. All will give and receive help to and from each other in accomplishing

whatever objective the team is working towards accomplishing. Support should be demonstrated daily by assisting volunteer members in their endeavors.

3. **Genuine communication.** Communication has two dimensions. The quality of openness and authenticity of the member who is speaking, and the quality of non-evaluating listening by other members. Open, authentic communication takes places when mutual trust and support are so well established that no member feels they have to be guarded or cautious about what they say. It also means that members of a good team won't "play games" with each other, such as asking "trap" questions or suggesting wrong answers to test another member's integrity. Non-evaluating listening simply means listening with "bias filters" removed to what the other person is trying to communicate. Most persons listen through an evaluative screen and tend to hear only those aspects of a communication which do not threaten status, roles or convictions.

4. **Accepting conflicts as normal and working through them.** Individuals differ uniquely from one another and will not agree on many things. An unproductive heritage left by the old school of "human relations" is the notion that people should strive for harmony at all costs. A good team (where mutual trust, mutual support and genuine communication are well established) accepts conflict as normal, natural and as an asset, since it is from conflict that most growth and innovation are derived. It is also worth noting that conflict resolution is a group process, and the notion that a manager can resolve a conflict between or among subordinates is a myth.

5. **Mutual respect for individual difference.** There are decisions which, in a goal-oriented organization, must be department decisions because they require the commitment of most or all of the members and cannot be implemented without this commitment. However, a good department will not demand unnecessary conformity of its volunteers. It is easy for a group to drift into the practice of making decisions for, or forcing decisions on, an individual where clearly, for their own growth and for the good of the organization, they should make the decision. The individual member should be free to ask advice from other members who, in turn, will recognize that they are not obligated to take the advice. A good organization delegates within itself. In a well-established organization, with high mutual trust and support, the leader, or any member, will be able to make a decision which commits the team. In such a team, only important issues need to be "worked through," and there is much delegation from leader to members, from members to members and even from members to leader.

Not only is it imperative that we motivate individual members within our organization, it is also important to motivate the group. There are six basic methods of motivating a volunteer force:

- provide a challenge
- provide quality training
- develop a team concept
- tell the membership they can't (imply they don't have the skill or expertise)
- make them special
- provide group recognition

A good team (where mutual trust, mutual support and genuine communication are well established) accepts conflict as normal, natural and as an asset, since it is from conflict that most growth and innovation are derived.

One of the primary methods of motivating a group is to provide them with a challenge. Challenges in a volunteer force come in many forms. They might include, but not necessarily be limited to, building projects, participation in special activities, fund raisers, program development, or any other item which taxes the group's recourses and requires them to take action. A second means of motivating a volunteer force is to provide quality training. Training which requires members to work as a team seems to accelerate the ability to motivate a group. Company evolutions, multi-company operations, and specific tasks which require a team to work to accomplish given results stimulate a volunteer force. Not only can we challenge our volunteer force through training, we can also provide members the opportunity to develop and present various components within the training program. The team concept can be developed by building in the need to work together to accomplish specific objectives. Another excellent method of developing the team concept is to incorporate several groups into a major objective or project, requiring specific members to work with each other in completing various aspects of the project, making them dependent on each other to accomplish the overall objective.

Telling volunteers they're not capable or they can't do something is an excellent means of motivating volunteers to act. Most groups do not like to hear negative thoughts about their ability. Many times results can be accomplished by merely stating that there is some doubt as to the ability of the group to perform.

Volunteers like to feel important. Volunteers like to feel special. A very good method of motivating a volunteer force is to provide them with special recognition and special treatment. This can be accomplished by isolating specific groups and/or the force as a whole and providing them with recognition for various tasks or projects they accomplish. A picture in a newspaper, a special word over the radio, or coverage on TV goes a long way in making your force feel appreciated.

One of the primary methods of motivating a group is to provide them with a challenge. Challenges in a volunteer force come in many forms.

Volunteers like to feel important. Volunteers like to feel special. A very good method of motivating a volunteer force is to provide them with special recognition and special treatment.

"Turn-offs" for your volunteer firefighters:

1. Poor communication
2. Not appreciated
3. Lack of recognition
4. Favoritism
5. Lack of responsibilities
6. Absence of standards
7. Lack of support
8. Lack of programmed responses
9. Paid members' attitudes
10. Poor training

"Turn-ons" for volunteer firefighters

1. Appreciation of job well done
2. Recognition and reward
3. Personal satisfaction
4. Challenging work

5. Good training

6. Alarm activity

7. Support from administration and paid personnel

8. New members

9. Positive attitudes

10. Doing a good job

11. Adequate funding

12. Good equipment

13. Meaningful incentives

Volunteer incentive programs

A primary way to maintain volunteers once their basic needs are met is to provide incentives. The primary incentive for volunteer firefighters is recognition and reward. This is accomplished through training, achieving organizational goals and objectives and meeting individual needs. In order for training and achievement to be meaningful, it must be recognized and rewarded.

There are many types of incentives utilized within the fire service to recognize volunteers for achievement. The key is not so much what the incentive includes or entails, but more important, how meaningful it is to the volunteer. Just as important as individual incentives are those incentives which can be offered to the group. These might include social functions such as summer picnics, Christmas parties, awards banquets, etc., or such items as monetary payment, reimbursement for expenses, certificates and awards which can be given to members for recognition, contribution, and participation. The list of incentives is as long as is a person's ability to be creative. The important factor in determining what incentives shall be utilized within a specific organization is to fit the incentives to the membership. What might be a valuable incentive to one group may be meaningless to another. In order for an incentive to work as a tool in assisting us to motivate and maintain volunteers, it must be meaningful.

There are many types of incentives utilized within the fire service to recognize volunteers for achievement. The key is not so much what the incentive includes or entails, but more important, how meaningful it is to the volunteer.

Compensating volunteers

The idea of compensating volunteers for participation has been around for a long time. The three primary means of compensating volunteers are through out-of-pocket reimbursements, flat hourly rate, and a point system. Out-of-pocket reimbursement compensates volunteers for actual expenses. Such reimbursement makes it necessary for the volunteer to complete an IRS form 990 yearly. Normally expenses include:

- fuel

- clothing

- educational materials

- meals

- safety equipment

- training expenses

- other volunteer related expenses

Flat rate reimbursement or "paid-on-call" volunteers are paid a specific dollar amount for tasks such as training or emergencies. This amount can be as low as $1.00-$2.00 per function or as high as $26 – $28 per function. Often rank determines the amount paid.

Note: The Fair Labor Standards Act has eliminated the ability of most volunteer organizations to pay an hourly rate.

Point systems are more complex and vary between departments. A point system can be based on the number of participants, value of a specific job, or rank. One advantage of a point system is the ability to weight activities. This encourages participation in key areas and serves as an incentive to increase participation in a specific activity. Point systems can be utilized in conjunction with a length of service awards program (LOSAP).

There is NO evidence that compensation positively or negatively affects a department's ability to recruit, train, or maintain active volunteers. In addition, we have not seen evidence that compensation affects the quality of service provided. Therefore, our suggestion would be if you are currently compensating volunteers – continue. To change the practice would be difficult and cause significant economical and psychological problems for your members. If you are not currently compensating your volunteers – don't start!

Incentives for volunteers

Free or Inexpensive Incentives

Volunteer of the month/quarter/year	Tell people they're missed while gone
"Thank You" notes or cards	Use of computer
Atta-Boy notes	Send get well cards, birthday cards, etc.
A simple verbal "Thank You"	Discounts in stores
Pat on the back	Good quality training
Visit with volunteers	Job rotation opportunities
A smile	Patches/badges/decals
Labeled area to place equipment	Probationary training
Recognizing length of service	Colored helmets by rank
Charts to show accomplishments	Sports activities
Participation in decisions	Fund raising functions
Suggestion box	Committee activities
Rotate leadership assignments	Establish rank structure
Profiles in newsletters/newspapers	Sponsorship in parades and fairs
Chart showing monetary value of volunteer efforts	Social hour after drill
Job opportunity board	Apparatus assignment
Performance evaluations	Accomplishment boards
Open communication	Personnel board with pictures
	Muster team

If you are currently compensating your volunteers – don't stop. If you are not currently compensating your volunteers – don't start! There is NO evidence that compensation positively or negatively affects a department's ability to recruit, train, or maintain active volunteers.

Put on public display

Keeping everyone informed

Holiday decorations

Use of facilities

Use of repair shop and tools

Pool party

Free ambulance service

Car lights

Credit union membership

Engine company of the year/quarter

Incentives Requiring Money

Awards banquet

Jackets/windbreakers/t-shirts

Baseball hats

Training opportunities

Uniforms

Safety Equipment

Annual picnic/parties

Service awards for 5-10-15-20 years

Injury insurance

Length of service program

Scanners/pagers

Travel expenses to attend meetings

New apparatus or equipment

Expense reimbursement

Pizza feeds

Safety awards

Placards on trucks

Subscriptions to trade journals

Trip to National Fire Academy

Multi-agency competition

Send to special schools

Beer mugs/coffee cups with name

Employee assistance

Good training facilities

Spouse appreciation dinner

Department newsletter

College reimbursement

Athletic event tickets

Pop machine

VCR/movies in station

Vehicle for officers

License plate frames

Gas money

Equipment bags

Retirees Hall of Fame

Video club membership

Gift certificates

Physical exams

$50.00 for newborn babies

4th of July fireworks

Easter egg hunt

Billboards with names showing appreciation

Weight room

Remember –
Fit the
incentive to the
needs and
desires of the
group!

Remember – Fit the incentive to the needs and desires of the group!

The rise and fall of the volunteer

Not until recently has much research been done regarding the rise and fall of volunteer firefighters, and what has traditionally been referred to as "volunteer burnout." A volunteer will normally reach a plateau at a level where they and their families are comfortable. This will be an activity level which provides the volunteer with the participation that was expected and desired, but yet doesn't interfere with the rest of their life to the point where it causes significant problems. This plateauing normally occurs three (3) to four (4) months after the recruit graduates and usually continues for approximately nine (9) to twelve (12) months. At the end of a year's period (give or take a few months) a volunteer will usually start declining in performance level. This will be noticed particularly in drill and alarm attendance. (This is one reason why comprehensive records are imperative.) The person responsible for individuals within a program will need to take necessary steps to bring the volunteer's performance level back in line when this decline occurs. We have found that when a volunteer or group of volunteers begin declining in participation, the program and/or the individual is in need of a new challenge or additional motivation. Challenge and motivation can usually be injected by providing members with new responsibilities and/or tasks for which they can be responsible.

Specialized training, sending a person to a workshop or seminar, or giving them new or increased responsibilities within the department usually work in providing a "shot in the arm" to the individual or group. Once this is done, records and data should show increased performance to the point where a plateau is once again met. This decline will show up again in another nine (9) to twelve (12) month period and will need to be dealt with just as it was a year earlier.

At this time you are probably wondering if this continual rise and fall is avoidable and if so, how it can be avoided. The answer, at least to our knowledge, is that it is not. Individuals and groups will continually have their ups and downs. The genius of persons responsible for maintaining performance levels within a volunteer department is to monitor the rise and fall and inject new ideas, concepts, and programs, etc., to keep individuals interested and enthused. If measures which are designed to increase proficiency and performance within the department fail, then the department is faced with what is considered "volunteer burn-out." Even if revitalization measures fail, counseling should be conducted prior to giving up on a volunteer. Short leaves of absences or other alternatives need to be pursued. The primary goal is to avoid totally losing the volunteer.

Not only do they have a lot of time and effort invested in the fire department, but you will have a lot of time, money, and effort invested in them. As is the case with so many situations involving volunteers: "An ounce of prevention is worth a pound of cure."

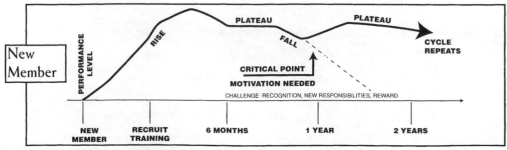

Make your program meaningful

A primary key when developing programs to assist in attracting, retaining, and maintaining volunteers is to make sure the program developed is meaningful to the volunteer group. It is the obligation of the organization's leadership to make sure that programs are designed to meet both short and long term needs of the membership. These concepts must be continually monitored when developing retirement programs for volunteers.

Fifteen years ago, the concept of retirement benefits for volunteer firefighters was alien to the majority of fire and rescue service personnel. While some states such as Minnesota, Oklahoma, and Texas encouraged retirement programs for volunteers, many such as New York, Pennsylvania, New Jersey, and Florida were unaware of the availability of the programs. In fact, while the fire service leadership in some states aggressively promoted legislation authorizing programs, most firefighters never took advantage of retirement programs. Finally, some fire service leaders felt that any benefits paid to volunteers would destroy the whole foundation of volunteerism.

Today, the large majority of fire service leaders recognize the value of retirement programs in attracting and, more importantly, retaining volunteer firefighters. Of the states listed above, where programs were nonexistent fifteen years ago, every one has studied the concept of retirement programs for volunteers and when needed introduced legislation authorizing the programs. Several years ago Delaware, Indiana and Connecticut joined the growing list of states which have adopted enabling legislation.

How do retirement programs for volunteers work? A simple question, but one which needs to be answered in a number of different ways.

Length of service awards program

In most areas surveyed retirement programs for volunteer firefighters are called Length of Service Awards Programs, LOSAP's, or Awards Programs. These terms are used to distinguish them from the type of retirement plan or pension plan employees are provided by their employers. While there are technical reasons for the distinction, the psychological reasons are equally important. Namely, that a Length of Service Awards Program is a plan whereby benefits are earned. They are not a give-away program and members do not earn benefits just by paying dues and putting in time.

Who is eligible

Eligibility is either determined by state statute or by the local plan sponsor. If you have a statewide plan, your first step would be to contact the person or agency which administers the program to determine the eligibility criteria. If you live in an area which does not have a statewide plan, you can develop eligibility standards which are customized to your needs or are contained in enabling legislation. These standards are often called a point system. A point system is a set of activities which must be accomplished by the volunteer in order to be eligible for the benefits of a Length of Service Awards Program. The point system should be developed to increase participation in areas of greatest need and be easily understood by the volunteers. The best way to develop a workable point system is to establish a committee to determine what areas are important to the fire department, what is considered an acceptable level of achievement and finally, must a firefighter earn points in that area.

For example:

Attendance at meetings

1. What is the average number of meetings a department has during a year? – 12

2. How many meetings does an average firefighter attend? – 6

3. How many meetings can a firefighter reasonably be expected to attend? – 8

4. Should attendance at a minimum number of meetings be mandatory? – Yes

Using the above criteria, the point system might read:

Attendance at meetings:

One point per meeting, minimum of 8 points must be earned from this category. No more than 12 points may be earned from this category.

Training

1. How many hours of training can a volunteer be reasonably expected to attend? – 60

2. What is the minimum acceptable number of training hours? – 24

3. What, if any, courses should be mandatory for new recruits? – Firefighting I

One point per training hour, a minimum of 24 points must be earned in this category including the successful completion of Firefighting 1 for all first year members.

The same types of questions should be asked for drills, emergency responses, non-emergency activities and any other activities important to the proper management of your fire department.

What benefits should be included?

Retirement benefit

The most important long term benefit is the monthly income payable at retirement. Generally, a portion of a member's retirement benefit is earned for each year he meets the eligibility criteria.

For example, a plan might have a benefit formula which states: "A member will earn $10 per month at retirement for each year of service." If a member is active for 10 years, the benefit payable at retirement would be $100 ($10 per year times 10 years). If the member remained active for 20 years, his benefit would be $200 ($10 per year times 20 years).

The benefit formula is one of three factors in the retirement benefit that directly and immediately affect the cost of the program. The greater the benefit formula, the more expensive the program. The other two areas are: crediting past service and the maximum benefit. Most plans establish a maximum retirement benefit for their program. For instance, in the example previously cited, the maximum benefit might be $300 per month at retirement. Therefore, 30 years is the maximum amount of service credited. The reason for the cap is two-fold. First, budget considerations, and second, because many fire service leaders believe that if you secure 20 or 30 good years out of a firefighter, you shouldn't expect more. The final factor which directly impacts cost is crediting past service (i.e., service before the plan began). If money was no object, all plans would probably credit all past service. Unfortunately, due to budgetary constraints, many departments limit the amount of past service to 20, 10 or 5 years. Other departments only award past service if a firefighter earns future service. For example, Emmaus Volunteer Fire Department in Pennsylvania credits three years of past service for each year of future service. So if a firefighter started

The most important long term benefit is the monthly income payable at retirement.

with the department 21 years ago, he would have to stay active for 7 additional years to earn all his past service.

After a member is active a number of years, he is vested. Vesting means ownership, or when the benefit is guaranteed. For example, if your vesting schedule reads 0-5 years – 0%, over 5 years – 100% vested, a member who quits after 4 years would lose all rights to his benefit. Conversely, a member who is active for over 5 years would be entitled to his earned benefit payable at his normal retirement age.

Disability benefit

In developing your Length of Service Award Program, consideration must be given to disability benefits. Questions asked should include: What is the definition of disability? Should it be limited to fire service activities? Should benefits begin immediately if disability occurs? What level of benefit should be paid? Many of these answers are based on other benefits available to your volunteers. For example, if you have strong workers' compensation benefits and/or accident and sickness coverage, your firefighters may be adequately covered for fire service related disabilities. On the other hand, virtually no insurance or benefits are available for non-fire related disabilities. Unfortunately, providing coverage for these disabilities on a self-insured basis could literally bankrupt a plan.

One alternative commonly used has no direct impact on the plan's assets. It is considered fair and while not exceptionally liberal, does ensure that the member receives his fair and equitable benefit from the plan. The suggestion is to use the Social Security definition, i.e., disabled for six months from any cause and unable to perform the duties of any occupation for which he/she is trained. Then define the benefit as the present value of benefit paid cash lump sum. This means the disabled member will receive the value of his earned benefit cash lump sum.

Preretirement death benefits

Most fire and rescue service leaders agree that if a member in good faith has been working hard to earn his retirement benefit, but unfortunately dies before receiving it, his family should receive a benefit. The question then becomes how much and what for? Part of the decision is based again on what other death benefits are payable to the firefighters. Generally, other benefits do a reasonable job of providing coverage for fire service related deaths, except for heart and circulatory deaths which might not be able to be traced to the fire or rescue activity. Therefore, in most cases, it's more important to focus on non-fire related deaths. Preretirement death benefits can and should be insured. The cost for insuring a preretirement 24-hour death benefit is insignificant when compared to the alternative of self insuring the benefit. For example, if you choose to self insure your preretirement death benefit, you would have two choices. First, provide benefits only in an amount equal to the member's earned benefit. The pitfall with this approval is the minimal amount of benefits payable on behalf of the younger firefighters. As chief officer of a fire or rescue department, could you imagine delivering a check to a firefighter's widow for $100 or $200? The alternative of providing a meaningful benefit of $10,000 to $25,000 could bankrupt the plan.

In developing your Length of Service Award Program, consideration must be given to disability benefits.

Most fire and rescue service leaders agree that if a member in good faith has been working hard to earn his retirement benefit, but unfortunately dies before receiving it, his family should receive a benefit.

By insuring a death benefit for a reasonable amount, you have accomplished three objectives:

1. You are providing a meaningful valuable benefit particularly to younger firefighters with families. Recognizing that retirement is too distant a goal for most young firefighters, the death benefit protection becomes even more important to the firefighter than their retirement benefit.

2. You are insuring or protecting your pension plan assets by not requiring withdrawals for deaths.

3. You are providing an additional disability benefit through the inclusion of a "waiver of premium" feature. Once a person is declared disabled, their life policy will remain in force for as long as the disability lasts with no premiums required.

When are benefits paid?

Benefits begin at the plan's normal retirement date or entitlement age. From a budgetary point of view, you can expect your cost to increase 6% to 10% for each year you lower the retirement age. For example, if the recommended deposit for your plan is $10,000 with an age 65 retirement, if you lower your retirement age to 62 you could expect your deposit to increase to between $11,800 and $13,000.

Other factors to consider when establishing a retirement age are: Should there be a service or plan participation requirement? For example, a retirement date might be stated as age 65 with 5 years of plan participation. This means that even if a member is already age 65 when the plan begins, they must remain active for at least five years. Or as a variation they might be able to retire but won't begin drawing benefits for five years.

How much does it cost?

As illustrated by all the variations discussed above, it is very difficult to clearly state what a Length of Service Awards Program would cost. It is also important to recognize that unlike typical insurance programs, the assets deposited to fund the program remain assets of the plan sponsor, subject to the various plan provisions. Therefore, it is best to consider contributions to a LOSAP as a prefunding of a future promise, not as an insurance premium.

The following are some of the typical factors that affect the deposit recommendation:

Ages — The longer you have to invest the funds, the lower the deposit recommendation. For example, a 50 year old's deposit might be $400. Whereas a 40 year old's with the same benefit might be $180.

Past service — The greater the credited past services, the higher the ultimate retirement benefit.

Benefit formula — The larger the benefit formula, the larger the benefits, the higher the recommended deposit.

Retirement age — The sooner a member can draw benefits, the less time to make contributions, the higher the contribution.

Ancillary benefits — Death and disability benefits will probably initially increase the deposit recommendation.

Assumed Interest — In determining how much should be contributed on behalf of each member the actuary (the person who determines contribution) assumes a rate of interest. This rate is the percentage which the actuary conservatively feels can be earned over the plan's lifetime, typically 40 years. Usually, this rate is between 5% and 7.5% based on the size of the plan. It is extremely important to realize that while your investments may currently be 7% or 8%, they can't be prudently expected to remain that high over the life of the plan.

Funding method

Actuaries use various methods to determine the proper amount of funds needed annually to provide the level of benefits. Some funding methods are developed to provide level deposits while others have lower initial deposits which increase in later years.

Comparing proposals

If you are planning to review proposals from several vendors, it is important that each vendor be given identical bid specifications. In addition to the benefit formula, retirement age, vesting, preretirement death and disability benefits, all vendors should use the same actuarial assumptions. Typically a bid specification should include the assumed interest rate, funding method, and any assumed turnover. If all vendors have the same plan, level information, and use the same assumptions , contribution requirements (exclusive of the preretirement death benefits) should be approximately equal. The selection process should consider experience, investment results, and administrative fees.

Supplement accident and sickness programs

Many fire districts, municipalities and fire departments secure accident and sickness insurance for their members. In states where there is no workers' compensation, these programs may be all the insurance a firefighter can depend on. In states with workers' compensation, these programs fill any gaps left by workers' compensation and provide additional benefits to show the community' s gratitude.

Usually all active members of the department are included automatically in the program. Coverage can and should be considered for junior members, auxiliaries and board of fire commissioners.

Generally, the accident benefits are paid for any "Covered Activity." The term "covered activity" includes emergency responses and meetings, drills, training, parades and any other activities sponsored by the organization.

The sickness section of the policy generally includes heart and circulatory malfunctions as well as specified conditions and infectious diseases. Sickness coverage is typically limited to emergency duties.

Benefits usually include accidental death and dismemberment in amounts ranging from $25,000 to $150,000. Medical benefits are also included and are intended to supplement any workers' compensation or other benefits programs. The intent is to ensure that all fire service activities are covered.

Disability Income is the third major component of an accident and sickness program. Many carriers have developed a variety of methods to provide adequate income replacement to disabled members. One of the more innovative ways insurance companies ensure adequate benefits is to utilize a multi-tiered approach. During the first 30 days of disability, the disabled firefighter receives the amount selected in the policy, typically, $150 to $300 per week. Then if the member is still disabled, a three-step formula is calculated:

1. The member's predisability wage is determined, extra income for long hospital stay.

2. Benefits received from W.C. and other group disability plans are calculated.

3. The difference between what the member earned before their disability and what they are receiving from other insurance is payable to the disabled member.

For example: assume a member earns $600 per week and becomes disabled. The disability benefit purchased is $200. During the first 30 days of disability, this member would receive this $200. After 30 days, this plan would look at the member's income of $600 and compare it to benefits they are receiving (let's say $200 from W.C.). The plan would then pay the member $400.

The reason this approach makes sense is that a disability income plan should try to ensure that the disabled person does not suffer any economic loss as a result of a disability. If, after 60 days, a member is still disabled but is not suffering any loss of income, a minimum disability benefit is paid.

Most innovative plans pay the benefit for five years. If this disability is severe, the plan will respond by also paying a cash lump sum benefit based on the degree of permanent impairment. For serious impairments, the benefit could be as high as $150,000. These funds allow the impaired person the financial freedom to adjust to their new life style.

Most progressive insurance policies also include benefits for vision impairment and burn disfigurement as well as supplemental income for extended hospital stays.

Group life benefits

In many areas, it is common for a fire district to issue all members a 24-hour all-cause life insurance benefit. Amounts range from $5,000 to $50,000 and are paid for deaths from any cause. The negative aspect of these programs is benefits often reduce or terminate at age 70.

Notes ■

Effective Leadership

12

From Firefighter To Officer

Being an effective leader is one of the most challenging and difficult tasks a fire officer faces. There is not a simple set of rules to apply to a situation that will be a cure-all. Every situation is different. A comprehensive knowledge of supervision is required for effective leadership.

The first day a firefighter becomes a supervisor their role changes. Instead of taking orders and direction, they are the person responsible for giving orders and direction. From that day on, the supervisor is faced with tasks, problems, and situations requiring well-developed decision making skills.

The Supervisor Looks At Their Job

Daily duties and responsibilities

- Getting the right person on the job at the right time
- Being cost conscious
- Attendance control
- Accident prevention
- Maintaining morale
- Adjusting complaints and grievances
- Improving discipline
- Keeping records and making reports
- Improving quality and quantity of work
- Planning and scheduling work
- Training personnel
- Coordinating resources
- Inspection, care, and maintenance of tools and equipment
- Work assignments
- Checking and inspecting work
- Settling differences between employees
- Encouraging team work
- Explaining rules, regulations, and policies

Major duties and responsibilities

- Plan and organize work
- Delegate responsibility and authority to get work done
- Supervise to get positive results
- Safety of employees
- Cooperation, development and maintenance

- Morale development and maintenance
- Training (greatest contribution)
- Records and reports

Common mistakes of new supervisors

It is safe to assume that all new supervisors will make mistakes. The newly appointed fire officer is no exception. Listed below are some of the more common mistakes made by new supervisors:

- New broom technique
- Making promises
- Dictatorial practices – orders vs. asking
- Playing favorites
- Careless remarks
- Failure to delegate
- Not accepting responsibility for actions
- Temper control when frustrated or under pressure
- Special privileges
- Becomes the person in the middle
- Allows the environment to become confused or appear to be disorganized

Common causes of supervisory breakdowns.

There are several common causes of supervisory breakdown. If this breakdown is avoided the overall effect of the supervisor can be dramatically improved. Some common causes of supervisory breakdown are listed:

- poor communication
- poor timing
- resistance to change (check attitudes)
- lack of clear cut objectives
- lack of policy, procedures, or practices
- morale deficiencies

The art of delegation

One of the most important tasks of any supervisor is to develop the art of delegation. It is imperative the supervisor be able to delegate to subordinates if they are to maximize the resources available and get the desired results in a timely fashion. Listed are some keys to proper delegation:

- be responsible for the delegation
- delegate to the lowest level possible
- delegate as much as possible

Reasons given for not delegating

- Lack of experience in subordinates. I'd rather do it myself than take the time to delegate.

- I can't take the risk of delegating, it's my neck.

- They expect me to do it.

- They don't see the "big picture."

- They don't want to do it, so they wouldn't do a good job.

Three types of supervisors

There are basically three types of supervisors. Listed below are the types, characteristics, and results one can expect from each supervisory type:

Autocratic — Characteristics of an autocratic supervisory style:

- dominates through rank

- needs power

- lacks consideration

Results expected from an autocratic supervisor style:

- poor morale

- poor work record

- grievances

Easy-Going or Democratic — Characteristics of a easy-going supervisory style:

- desires to please

- trusts people at all costs

- fails to correct deficiencies

Results expected from an easy-going supervisor style:

- poor morale

- sloppy work

- real leaders develop in group

- may eventually become autocratic

Effective or Situational — Characteristics of a effective supervisory style:

- is assertive

- uses leadership skills and methods

- uses personnel to assist

- situational in style

Results expected from an effective supervisory style:

- teamwork

- cooperation

- trust

- high morale

Becoming a better supervisor

It has been said that the key to any organization's success lies in getting the job done and the supervisor is the key person. Fire service officers face unique problems in dealing with subordinates and accomplishing the mission.

There are numerous methods of solving supervisory problems. However, three methods that appear to have outstanding merit are:

- professionalism
- integrity
- maturity

Professionalism

It usually pays to seek out professionals and learn from them – watch them in action and work with them. Remember, people in your community like to work with professional people. Without exception, it is the pro who has the respect and admiration of their fellow workers, as well as others.

Who is a "pro" in the fire service? More often than not, a pro in the fire service is a person who:

- is willing to do more than they are paid to do
- has a good work attitude
- is willing to take risks in trying new ideas and techniques
- is in complete mastery of the job
- has a personal sense of responsibility and activity
- seeks out problems and solves them
- coaches others

Check the above list. If you do these things, then perhaps you are a pro. If you don't, you run numerous risks. You and your company may become stale. You may begin giving in, rather than standing your ground on issues you feel strongly about.

Don't play follow the leader. Don't postpone decisions. A true professional is a strong dynamic person capable of dealing with responsibility.

Integrity

The power and authority given to fire officers by the people in the community demands integrity of the highest type. If this integrity is not maintained, then eventually authority will be abused. Only through a sense of morality and integrity can we gain the respect and confidence of others.

Maturity

Many supervisors are effective because they are mature individuals. The development of maturity should be a permanent goal. It is the mature person who gains the confidence of their subordinates. Ten characteristics of a mature individual are:

1. **Be willing to accept your self.** The mature officer is aware of their strengths and weaknesses. They know they are able to make the most of what they have and are free from the frustration of trying to be what they are not.

2. **Respect others.** The mature officer must know the strengths, weaknesses, abilities, and deficiencies of others. If they are mature, they have respect for others'

Don't play follow the leader. Don't postpone decisions. A true professional is a strong dynamic person capable of dealing with responsibility.

strengths and weaknesses. In other words, they have the ability to accept people as they are, not as they might wish them to be.

3. **Accept responsibility.** The mature officer can recognize responsibility and is willing to accept it even in situations which might not be of their making. They can recognize that people need strength and someone they can lean on. Realize it is the leader's responsibility to provide this strength. The immature person will complain long and loud about their place in the department and in life. They will never accept responsibility for their deficiencies, but instead will try to blame others for their problems.

4. **Be confident.** The mature officer is confident of their ability. They welcome participation by others in the area of decision making. They gain satisfaction through the accomplishments of their people. They help them to grow and whenever possible, step aside and allow them to assume responsibilities. They find a sense of personal gain and satisfaction in contributing to the development of other people.

5. **Be patient.** The mature officer accepts the fact that some problems have no easy answers or solutions. They do not accept the first obvious answer, but gather the facts and all available information. They are patient in seeking solutions to problems and in working with superiors.

6. **Make decisions.** The mature officer can make decisions in spite of ambiguities. They recognize that there is a time when action must be taken, when indecision is equivalent to making a decision not to act. In regard to the future, there can never be certainty – chances must be taken in fire situations, as well as in day-to-day department operations.

7. **Be resilient.** The mature officer is the person who bounces back after failure or hurt. They know that some failures or hurts cannot be escaped. They accept these things as part of life itself. They do not have to pretend that they like their failures, nor do they have to hide their hurts. They do, however, have the ability to accept and learn from them.

8. **Welcome work.** A mature person welcomes work. Welcoming work simply means that the person enjoys what they are doing and receives satisfaction from it. It is rare to find a good supervisor who fears or dislikes their work.

9. **Have strong principles.** The officer who is willing to fight for what they believe in shows a real mark of maturity. Many officers look upon the fire department as a trust which has been temporarily granted to them to nurture, develop, and then pass on to others. Officers who believe in this principle can stand for their beliefs for the good of the fire service rather than for personal gain.

10. **Have a sense of proportion.** The mature officer who has perspective can live a balanced life. They must be able to work hard, but also be able to relax away from work.

Styles of behavior

	I The Kindly Helper	II The "Doer" Man of Action	III The Thinker
How they influence people	• Favors • Entertaining • Pity	• Orders • Threats • Depriving	• Logic – facts • Shrewd argument • Knowledge
What they fear	• Loss of affection • Attack	• Loss of power, e.g. illness, being a "softy"	• Confusion • Obligation • Feeling
Their typical jobs	• Personnel • Teaching • "Assistant to"	• Manufacturing • Promoter • Surgeon	• Scientist • Engineer • Accountant
What they do under stress	• Clinging dependency	• Impulsiveness, action at any cost • Domination	• Withdrawal • Rule-ridden
What do they need?	• Stand up for yourself • Learn to say "no" • Ask for things • To make critical evaluation of others and their ideas	• Patience • To give support to others • Planning, decrease impulsiveness • Ability to listen	• Awareness of feelings • Less rule bound • Accept their own needs for closeness to others; "No man is an island"

Maturity checklist

1. Do you willingly accept responsibility for your subordinates' mistakes?
   ☐ Yes ☐ No

2. Do you patiently listen to people with whom you disagree?
   ☐ Yes ☐ No

3. Do you have a system for obtaining facts, making decisions, and implementation?
   ☐ Yes ☐ No

4. Do you take unpopular stands on issues in which you believe?
   ☐ Yes ☐ No

5. Do you voluntarily accept new challenges and/or responsibilities?
   ☐ Yes ☐ No

6. Do you cheerfully share credit with colleagues who help you?
   ☐ Yes ☐ No

7. Do you re-examine assumptions you have always taken for granted?
   ☐ Yes ☐ No

8. Do you accept unfavorable feedback about yourself?
   ☐ Yes ☐ No

9. Do you feel confident that you can handle your problems?
   ☐ Yes ☐ No

10. Do people seek you out for advice and assistance?
    ☐ Yes ☐ No

Leading others — managing yourself

Organizational structure is critical to the fire service. However, there are advantages and disadvantages of such a hierarchy.

1. Accept people as they are, not as you would like them to be. This can be seen as the height of wisdom – to "enter the skin" of someone else, to understand what other people are like on their terms rather than judging them.

2. Approach relationships and problems in terms of the present, rather than the past. It is true we can learn from past mistakes. However, using the present situation, not how it was in the past, will help you be a more productive leader. Dealing with the present is more psychologically sound than rehashing the past.

3. Treat those who are close to you with the same courteous attention that you extend to strangers and casual acquaintances. This skill is often most obviously lacking in our family relationships, but is equally important at work. We tend to take those we are closest to for granted. We get so accustomed to seeing and hearing them that we lose our ability to listen to what they are really saying or to appreciate the quality, good or bad, of what they are doing. We allow personal feelings of friendship, hostility, or differences to interfere with our ability to show courteous attention.

There are two aspects to this problem of over familiarity. First, not hearing what is being said, or selective deafness, will lead to misunderstandings, misconceptions, and mistakes. The second concerns feedback. Failing to give appropriate feedback indicates a low level of attentiveness.

4. Trust others, even if the risk seems great. Withholding trust is often necessary for self-protection but the price is too high if it means always being on guard and suspicious of others. An overdose of trust that at times involves the risk of being decided or disappointed is wiser than assuming that most people are incompetent, insincere, or untrustworthy.

5. Realize it is not necessary to receive constant approval and recognition from others. The need for constant approval in a work situation can be counter-productive. It should not matter how many people like their leaders. The important thing is the quality of work that results from collaboration. The emotionally wise leader realizes that work quality will suffer when undue emphasis is placed on being a "good guy." It is more important to realize that a large part of the leader's job is to take risks. Risks, by their very nature, cannot be pleasing to everyone.

Leadership style

The basic concept of leadership is based upon three ways of running a fire department:

- authoritarian
- democratic
- participative

Which approach is most effective will change with the situation. The successful leader recognizes the nature of the problem they are dealing with then they adopt the appropriate style of leadership.

Definition of leadership

- Leadership is the art of dealing with people in a manner that commands their respect, support, and cooperation.

- Leadership is also the activity of influencing people to cooperate toward some goal which is desirable.

- Additionally, leadership is getting people to do something because they want to.

Possibly one of the best definitions of leadership is as follows: *"Getting things done through people – how you want it done, when you want it done, willingly. "* – Dwight D. Eisenhower

Approaches to effective leadership

The following statements from fire officers represent different attitudes and approaches to "effective leadership." They illustrate the dilemma in which the fire officers may find themselves.

- I believe in delegating problems to others and leaving it up to them to solve the problem. I serve only as a catalyst, reflecting their thoughts and feelings so they can better understand them.

- I don't believe in making a decision myself on matters that affect other people. I believe in talking things over with other officers or subordinates. However, I make it understood that I'm the one who will have the final say.

- If I decide on a course of action I feel I should do my best to sell the course of action to those affected.

- I've been selected as the officer/chief. It's my responsibility to make the decision. If I let others do it, then I'm not needed.

- If you want to get something done, it's best to do it yourself. I don't waste time on meetings or trying to get opinions from others. Someone has to be responsible and since I'm the officer, it should be me.

As you can see, these styles travel from one extreme, that of being a "democratic" leader, to the other extreme, being an "autocratic" leader. We maintain one should be a situational leader, which means your leadership style changes to fit the situation. More importantly, your leadership style should fit your volunteer force. Remember, you are not as good somebody else, but you are the best "you" in the world. Be yourself.

Indicators of leadership

High morale, measured through:

    a. persons volunteering to do more

    b. good work results

    c. participating in special activities

Discipline, measured through:

    a. number and type of non-desirable performances

    b. number and type of grievances

    c. number and type of positive reports by public

Proficiency, or high degree of knowledge and skill, measured through:

    a. job results, both emergency and non-emergency

    b. performance evaluation results

    c. feedback – users and observers

Mutual respect and confidence among people, measured through:

    a. verbal communications

    b. written requests related to work assignments

Dependability, measured through:

    a. work results

    b. work record — sick leave, tardiness, etc.

Four keys to becoming a leader

The four keys to becoming a strong leader are:

- Develop a vision toward the future. You must know where you are headed and where the organization needs to go.

- Exhibit adequate oral and written communication skills.

- Positioning, which involves being in the right place at the right time.

- Set the example. Your people need to be able to look to you for direction and guidance. Do a self test to evaluate what type of an example you are setting by considering the following statement: If someone came to your department and asked your people, "When the chips are down and the going is tough, who would you want to follow through that difficult time?" would your people respond with your name?

Tactics for leadership success

- Loyalty to those above and below you in rank.

- Maintain a strong work ethic. Be responsible by doing what you say you will do.

- Avoid losers — negative thinking carries over to those that come in contact with you.

- Put salve on your wounds. It is not easy to be out in front leading. You will be hurt emotionally at times. Acknowledge the hurt and deal with your feelings.

- Calm troubled waters. Teamwork develops as people are allowed to participate in settling difficult situations.

Leadership attributes checklist

Bearing

    a. Carriage.

    b. Maintain a high level of alertness and energy.

    c. Personal conduct. Avoid coarse behavior.

    d. Require high dress standards for yourself.

    e. Be a good listener.

Physical and Mental Courage

a. Recognize danger and/or criticism.

b. Provide the ability to go ahead. **This is vital**.

c. Moral courage. Be able to stand up for what is right when it may be unpopular.

d. Understand your fears.

e. Control fear through self-discipline and calmness.

f. Accept the blame when you are at fault.

Be decisive

a. Make decisions promptly.

b. Announce decisions clearly and forcefully.

c. Exhibit positive actions.

d. Recheck your decisions.

e. Analyze decisions made by others.

f. Broaden your viewpoints by listening to others.

Be dependable

a. Provide the certainty of proper performance.

b. Don't make excuses.

c. Sign your name to tasks performed. If you find you don't want your name associated with the task, keep working on it until you can be proud of the task performed.

d. Be exact in details.

e. Be punctual.

f. Carry out the intent and spirit of orders and/or directions given.

Be able to endure

a. Your physical condition should allow you to withstand a reasonable amount of pain, fatigue, stress, and hardship.

b. Avoid nonessential activities that lower your stamina.

c. Test your endurance often.

d. Force yourself to go on.

Enthusiasm

a. Display sincere interest and zeal.

b. Understand your mission and believe in it.

c. Be optimistic.

d. Explain the "why" to your people.

Initiative

 a. Be a self-starter.

 b. Be alert.

 c. Be observant.

 d. Readily accept responsibility.

Integrity

 a. Develop sound character.

 b. Practice honesty and truthfulness.

 c. Be accurate.

 d. Place moral values and duty first.

Judgment

 a. Logically weigh facts and alternatives.

 b. Practice making estimates.

 c. Anticipate the need to make decisions.

 d. Approach problems first using common sense.

Knowledge

 a. Show professional competence.

 b. Maintain a professional library.

 c. Keep up with current events.

Note: Refer back to the "Leadership Attributes Check list." Do the following exercises:

 1. Underline your strengths.

 2. Place a star by your weaknesses.

 3. Develop a plan on how you can improve those areas which are currently weak.

Motivation — the key element in leadership skill

To motivate others a leader must:

- Show persistency in meeting objectives.

- Have an ability to communicate effectively.

- Show willingness to listen objectively.

- Have an understanding of their people and how they will react to situations.

- Treat everyone objectively.

- Display forthrightness.

Principles of supervision

 Along with learning and practicing the desirable qualities of supervision, the following principles will help you to do your job better:

- Volunteers must clearly understand what is expected of them.

- Poor work deserves constructive criticism.

- Good work should be recognized.

- Volunteers should be encouraged to improve themselves.

- Volunteers should be allowed to show that they can accept greater responsibility.

- Volunteers should work in a safe and healthy environment.

- Volunteers need guidance in their work.

Leadership — make things happen

Individual priorities have changed since the 1950's regarding work and family. In the early 1950's people held work as their number one priority with the family coming second. We are now dealing with people who have put the family as their number one priority and the work place second.

This fact does not minimize the importance of utilizing good leadership skills to assist personnel in establishing successful careers. Leadership skills will also help you to be successful in satisfying organization needs, such as goals and objectives.

Three basic human needs must be fulfilled for an individual to experience true job satisfaction:

1. People need to be appreciated for a job well done.

2. People need to have challenges available to experience a sense of accomplishment.

3. People need to belong to an organization that they can be proud of.

Superior leaders demonstrate the following characteristics:

- They listen and coach. Seventy percent of their communication time is spent listening. They ask more questions than they give directions. When coaching, most of their time is devoted to asking questions. They allow individual natural abilities to be used.

- A superior leader uses calm, flexible conflict resolution. Their number one strategy is problem solving. Problems have four basic roots:

  a. value differences

  b. perception differences

  c. little or no information

  d. conflict of egos

- Superior leaders have positive expectations. They think positively and believe that most people (around 90%) are okay.

- Superior leaders have a sense of humor.

- Superior leaders don't take themselves too seriously.

- A superior leader won't lie or cheat, but will take advantage of situations in order to get to where they believe an organization should be going. This must be done in an open and honest manner.

- Superior leaders set goals. Nothing will take place unless you have attainable goals. Accomplishments are not accidental.

- Superior leaders have a vision. Some examples of visions are:

   We are the best!

   We help!

   We treat people right!

   We have quality!

   We save lives!

   We prevent fires!

- Superior leaders represent their people well. They stand by their people. They do not degrade others.

- Superior leaders should mix with their people to understand their perspectives. This is accomplished only by spending time with them to develop a feeling of where they are.

Essentials of cooperation

Cooperation is essential to the development of a team. In order to achieve desired results as a fire department, cooperation must be learned, developed, and enhanced through the group. The essential elements of cooperation are:

- Trust

- Common goals

- Mutual interests

- Understand special interests of each group

- Communication

- Group interests come first

Achieving cooperation requires:

1. Knowledge of common purposes

2. Participation

3. Action

Attitudes

Attitudes are important to the well being of a group. Group attitudes will affect what the group is able to accomplish. The supervisor must understand the "condition" of the group and what will affect membership attitudes. Attitude has been defined as:

1. mental position regarding a fact or state of affairs

2. a feeling or emotion regarding a fact or state of affairs

Attitudes play an important role in the outcome of activities. Experts believe that up to 60% of a person's performance is based on their attitude and motivation toward what they are doing. It is important to know that it takes time and effort to change how a person feels and thinks about a given situation. It is even more difficult to change group attitude. We must begin with less difficult tasks and work up in order to change group attitude. Attitudes are influenced by:

- Values

- Social position

- Experience

- Current situation

An officer can set an example by resisting the negatives and promoting the positives. Officer attitude is influenced by subordinates. Subordinate attitude is influenced by officers. The officer has an advantage since subordinates usually want to please the officer.

Following are lists of personality characteristics people with positive and negative attitudes exhibit. You will find that the person you find most difficult to deal with is telling you something about yourself.

Characteristics of a positive attitude

- Gets things done

- Motivated

- Constructive criticisms

- Smiles

- Good outlook

- Good self-image

- Unselfish

- Happy

- Internal

- Flexible

Characteristics of a negative attitude

- Griper

- Complainer

- Procrastinator

- Frowns

- Poor outlook

- Selfish

- Sad

- External

- Rigid

There are a few advantages, or payoffs, to having a negative attitude. A few of these advantages are:

- Examine issues

- Gets attention

- Never disappointed

- Thrill of being right when things go bad

- Disguise incompetence

There must be positive and negative to be balanced. The "positive" person will create a natural vacuum and attract a negative. The negative person will try to be in control. Don't let them, you must handle the situation. The negative person will not stop on their own.

A practical, positive attitude is not the opposite of a negative attitude. Learn to use the good aspects of positive and negative attitudes to their advantages. Try taking a plan to a negative person and ask them to look it over and find everything that could go wrong. Tell them, "I know I'm being an optimist. You show me the things that can go wrong." Every department needs a balance of optimism and pessimism.

Handling a negative person

The following are steps you should take when dealing with negative attitudes:

1. Identify the source.

2. Bring into the open.

3. Evaluate "why."

4. Develop a "cure."

5. Be positive and assertive.

If a negative statement is made, simply respond with, "Well, you may be right." Then quickly move on to other matters. You have not given the negative person the attention they are looking for.

Let the negative person give their opinion and then say, "And what else?" Add more negatives, indicating it may be worse than they think. Be better at their game than they are. They may come back with a positive comment. When you agree with the negative, it may throw them off balance.

Periodically give the griper a little time. They deserve it – after all, it is their job. Demonstrate that you are willing to listen up to a point. In using this technique, you are using conflict in a positive manner. A moderate amount of conflict, or differing opinions, is essential. Your department will become stagnant if you maintain a policy of no conflict.

If the negative attitude increases, break the pattern. Be willing to confront the issue and take control. You may say, "You are getting so negative it's very difficult for us to be around you. Please back off." The most important point is that you stay in control of the situation.

Conflict resolution

Take the following steps to resolve conflict:

• Ask the person what is wrong.

• Listen to their answer.

• Repeat it back to make sure you understand what they said. Clarify. Ask if you repeated what they had stated correctly. Legitimize their feelings. Let them know they have a right to their feelings.

• Add to their case.

• Explain the problem or goal.

There are ways to argue fairly and unfairly. We call fighting unfairly "fighting dirty." Listed below are characteristics of fighting fairly and dirty:

Fighting fairly

- Ask the other person for permission to speak with them about something that has been bothering you.
- State your feeling and the triggering behavior.
- Ask about reasons and intentions.
- Ask for what you want.
- Discuss and develop a workable plan.
- Thank the person for listening regardless of the outcome.

Fighting dirty

- Catch the other person by surprise.
- Point out all of the other person's negative characteristics.
- Make accusations about motives.
- Write the person off.
- Refuse to talk.
- Observing and confronting conflict.
- Gossip behind their back.

Attempt to keep the conflict resolution process impersonal. Do not "tell" the person what they must do. Suggest what you feel would be the best route for them to take. You should spend most of your conflict resolution time in the "ask and listen" mode. Ask questions such as, "Can I count on you to… ?" or "What do you need from me to be successful at this?"

Self-critique

Answer these questions after dealing with conflict resolution:

- Did I observe others' behavior and attitudes to learn from them?
- How did I react to opinions expressed by others?
- Did I avoid behavior that could cause negative reactions in others?
- Was I able to maintain a friendly spirit among the group?
- Was I able to inject humor or compromise when appropriate?
- Was I open to suggestions?
- Were my ideas helpful? If not, why?

Observing and confronting conflict

Instructions to Observer: While observing two people talking, use the checklist below to observe how they interact. Don't pay attention to what they say. Make a check mark each time one or the other does one of the things listed below. Write in any other behaviors that you notice.

Person #1 Check Marks	Useful Actions	Person #2 Check Marks
_____	Listened first, listened attentively.	_____
_____	Asks other person to state views.	_____
_____	Repeats back. Re-phrases what was said. Checks to see if heard correctly.	_____
_____	Asks questions which draw the other person out.	_____
_____	Legitimizes the other's position. Says: "That makes sense." "I can see why you feel that way." "You are right about that."	_____
_____	Adds another argument which supports the other person's position.	_____

Actions That Maintain Conflict

Person #1		Person #2
_____	Takes over and presents own side first.	_____
_____	Dominates conversation. Works at selling ideas. Uses persuasion or logic to bring the other person around. Explains why their view is the right view.	_____
_____	Interrupts the other person. Scoffs, ridicules, or criticizes other person's points. Negative tone of voice.	_____
_____	Gets angry or irritated when other person rejects statements and holds to own.	_____
_____	Gives negative non-verbal feedback. Shakes head "No." Scowls. Looks away with negative facial expression.	_____
_____	Says "Yes, but..." after other person has made a point.	_____

Cooperative New Direction

Person #1		Person #2
_____	After carefully listening to other person, asks if there are any arguments in favor of the other side.	_____
_____	Describes own view of the problem to be solved and asks for suggestions about a better solution.	_____

Positive approaches to criticism

- Criticize where and when warranted.
- Criticism can be used to boost or destroy morale.
- Criticism should be constructive.
- Criticize the problem, not the person.
- Criticize only to the degree needed.
- Include the solution to the problem.
- Be sure of your facts.
- Have a positive for every negative.
- Avoid losing your temper.

Being assertive

Assertiveness is another key to being an effective leader. Taking a class in assertiveness training is like purchasing dynamite. Assertiveness can be utilized in constructive ways or in a manner which can cause harm to individuals and the group.

Being assertive means doing and saying what needs to be said and done! To be assertive, you must be organized and work toward goals. You must understand your purpose!

Being assertive implies:

- I count – you count (versus I count – you don't count).
- Two-way, face to face communication.
- Being honest.
- Respects rights of self and others.
- Honors written agreements or discusses agreements and moves to realign them.
- Self-enhancing, but not at the expense of others.

A few of the reasons for being assertive are:

- you are a valuable resource
- time is valuable
- commitment to carefully planned goals
- you are responsible
- you are accountable

We often say "yes" when what we want to say is "no." Some of the reasons people say "yes" are:

- habit
- scared
- guilt
- don't have a better idea
- don't have anything better to do
- like to please
- trade-off

Rights the assertive person claims

- I am my ultimate judge.
- I have the right to expect to be treated with respect.
- I have the right to be listened to.
- I have the right to say no.
- I have the right to ask for information.
- I have the right to question why, how, when, where.
- I have the right to say, "I don't know."
- I have the right to voice my views.

Ways to be more assertive

- Describe your reaction.
- Be specific rather than general.
- Express disapproval at the right time.
- Express yourself in the first person (I, me).
- Be honest.
- Wait for the right moment to talk. Show the same courtesy to others as you would like shown to you.

Observable differences between effective and ineffective work groups

Effective Groups

1. The "atmosphere," which can be sensed in a few minutes of observation, tends to be informal, comfortable, relaxed. There are no obvious tensions. It is a working atmosphere in which people are involved and interested. There are no signs of boredom.

2. There is a lot of discussion in which virtually everyone participates, but it remains pertinent to the task of the group. If the discussion gets off the subject, someone will bring it back in short order.

3. The task or the objective of the group is well understood and accepted by the members. There will have been free discussion of the objective at some point until it was formulated in such a way that the members of the group could commit themselves to it.

4. The members listen to each other! The discussion does not have the quality of jumping from one idea to another unrelated one. Every idea is given a hearing. People do not appear to be afraid of being foolish by putting forth a creative thought even if it seems fairly extreme.

5. There is disagreement. The group is comfortable with this and shows no signs of having to avoid conflict or to keep everything on a plane of sweetness and light. Disagreements are not suppressed or overridden by premature group action. The reasons are carefully examined, and the group seeks to resolve them rather than to dominate the dissenter.

On the other hand, there is no "tyranny of the minority." Individuals who disagree do not appear to be trying to dominate the group or to express hostility. Their disagreement is an expression of a genuine difference of opinion, and they expect a hearing in order that a solution may be found.

Sometimes there are basic disagreements which cannot be resolved. The group finds it possible to live with them, accepting them, but not permitting them to block its efforts. Under some conditions, action will be deferred to permit further study of an issue between the members. On other occasions, where the disagreement cannot be resolved and action is necessary, it will be taken but with open caution and recognition that the action may be subject to later reconsideration.

6. Most decisions are reached by a kind of consensus in which it is clear that everybody is in general agreement and willing to go along with their wishes in order to get results. However, there is little tendency for individuals who oppose the action to keep their opposition private and thus let an apparent consensus mask real disagreement. Formal voting is at a minimum; the group does not accept a simple majority as a proper basis for action.

7. Criticism is frequent, frank, and relatively comfortable. There is little evidence of personal attack, either openly or in a hidden fashion. The criticism has a constructive flavor in that it is oriented toward removing an obstacle that faces the group and prevents it from getting the job done.

8. People are free in expressing their feelings as well as their ideas both on the problem and on the group's operation. There is little pussyfooting, there are few "hidden agendas." Everybody appears to know quite well how everybody else feels about any matter under discussion.

9. When action is taken, clear assignments are made and accepted.

10. The chairman of the group does not dominate it, nor on the contrary, does the group defer unduly to them. In fact, as one observes the activity, it is clear that the leadership shifts from time to time, depending on the circumstances. Different members, because of their knowledge or experience, are in a position at various times to act as "resources" for the group. The members utilize them in this fashion and they occupy leadership roles while they are being used. There is little evidence of a struggle for power as the group operates. The issue is not who controls but how to get the job done.

11. The group is self-conscious about its own operations. Frequently, it will stop to examine how well it is doing or what may be interfering with its operation. The problem may be a matter of procedure, or it may be an individual whose behavior is interfering with the accomplishment of the group's objectives. Whatever it is, it gets open discussion until a solution is found.

Ineffective groups

1. The "atmosphere" is likely to reflect either indifference and boredom (people whispering to each other or carrying on side conversations, individuals who are obviously not involved, etc.) or tension (undercurrents of hostility and antagonism, stiffness and undue formality, etc.). The group is clearly not challenged by its task nor are they genuinely involved in it.

2. A few people tend to dominate the discussion. Often their contributions are way off the point. Little is done by anyone to keep the group clearly on track.

3. From the things which are said, it is difficult to understand what the group task is or what its objectives are. These may have been stated by the chair initially, but there is no evidence that the group either understands or accepts a common objective. On the contrary, it is usually evident that different people have different, private, and personal objectives which they are attempting to achieve in the group, and that these are often in conflict with each other and with the group's task.

4. People do not really listen to each other. Ideas are ignored and overridden. The discussion jumps around with little coherence and no sense of movement along a track. One gets the impression that there is much talking for effect – people make speeches which are obviously intended to impress someone else rather than being relevant to the task at hand.

Conversation with members after the meeting will reveal that they have failed to express ideas or feelings which they may have had for fear they would be criticized or regarded as silly. Some members feel that the leader or the other members are constantly making judgments of them in terms of evaluations of the contributions they make, and so they are extremely careful about what they say.

5. Disagreements are generally not dealt with effectively by the group. They may be completely suppressed by a leader who fears conflict. On the other hand, they may result in open warfare, the consequences of which is domination by one sub-group over another. They may be "resolved" by a vote in which a very small majority wins the day, and a large minority remains completely unconvinced.

There may be a "tyranny of the minority" in which an individual or small sub-group is so aggressive that the majority concedes to keep the peace or to get on with the task. In general only the more aggressive members get their ideas considered, because the less aggressive people tend either to keep quiet altogether or to give up after short, ineffectual attempts to be heard.

6. Actions are often taken prematurely before the real issues are either examined or resolved. There will be much grousing after the meeting by people who disliked the decision, but failed to speak up about it in the meeting itself. A simple majority is considered sufficient for action, and the minority is expected to go along. Most of the time, however, the minority remains resentful and uncommitted to the decision.

7. Action decisions tend to be unclear – no one really knows who is going to do what. Even when assignments of responsibility are made, there is often considerable doubt as to whether they will be carried out.

8. The leadership remains clearly with the committee chair. They may be weak or strong, but they sit always "at the head of the table."

9. Criticism may be present, but it is embarrassing and tension-producing. It often appears to involve personal hostility, and the members are uncomfortable with this and unable to cope with it. Criticism of ideas tends to be destructive. Sometimes every idea proposed will be "clobbered" by someone else. Then, no one is willing to stick their neck out.

10. Personal feelings are hidden rather than being out in the open. The general attitude of the group is that these are inappropriate for discussion and would be too explosive if brought out on the table.

11. The group tends to avoid any discussion of its own "maintenance." There is often much discussion after the meeting of what was wrong and why, but these matters are seldom brought up and considered within the meeting itself where they might be resolved.

Planning and organizing work

A supervisor's responsibility in planning is three-fold:

1. An obligation to the company to get work done quickly and economically.

2. An obligation to fellow supervisors to meet joint schedules.

3. An obligation to subordinates to make work efficient and rewarding.

What is meant by planning? Planning is the thinking that precedes actual job performance. It is an effort to answer two questions:

1. What is it that I have to do?

2. What is the best way to do it?

Unless these two questions are answered clearly and intelligently well in advance of actual task performance, the project may end in failure. Last minute planning is almost as damaging as no planning at all. Good planning is organized, proceeding step by step. An orderly plan will result in more problems solved, fewer trials necessary, fewer false starts, less time used, and less unnecessary work. Planning could proceed like this:

1. **What do I want to accomplish?**

   a. Define the objective.

   b. Make sure it is the right one.

   c. Against what standards will the job be measured?

2. **What are the ground rules?**

   a. What degree of freedom do I have?

   b. Who set the ground rules? Can they be bypassed or renegotiated?

3. **What data do I need?**

   a. Inventory of materials.

   b. Skills inventory of people available to do the job. How many will I need?

   c. Measure the job with a flow chart, flow diagram, blueprints, etc.

   d. What are the machine speeds, floor space, material needs, etc.?

   e. Consider past experience.

   f. What new equipment or techniques are available?

   g. What are the quality considerations?

4. How to proceed?

    a. This decision is based on an analysis of the data collected in Step 3.

    b. Consider each assignment by deciding:

- What is to be done?
- Who does it?
- Where and when it's done?
- How it should be done?

    c. Communicate and interpret this decision to all concerned.

    d. Assign responsibilities to people, machines, and space. Delegate the functions.

5. Test the plan.

    a. Get sufficient feedback.

    b. Can the machines turn out the work?

    c. Can the people perform as expected?

    d. Be flexible, expect to make changes and adjustments.

    e. Have an alternative plan ready if it is necessary.

6. Evaluation of the work.

    a. Have the objectives been met within the ground rules laid out?

What does it take to be a competent manager?

James L. Hayes groups management competencies into four generic categories:

1. action orientation
2. leadership skills
3. human relations skills
4. supervisory skills

Action Orientation – A good manager is someone who:

- is concerned with efficiency. This includes planning, setting priorities, and implementing them. Prioritizing is key. A good manager might know of 15 existing problems, but if they deal with all 15 the effort is dissipated. An effective manager selects no more than five critical things to work on at any one time.

- cuts through red tape. A good manager must keep things moving and is never satisfied with the speed at which things are moving.

- is concerned with impact and results. A good manager must constantly ask questions such as, "If we do X, what will be the impact on public relations?" A good manager can't get bogged down with process, but must be concerned with results. It doesn't matter what color Napoleon's horse was, but rather what was the outcome of the battle.

- is analytical and systematic. A good manager is always trying to put things in sequence. They try to find the hidden corners and to get a task to unfold in a systematic way.

Leadership Skills — A good manager:

- has self-confidence. A good manager is often wrong, but never in doubt. A hesitating leader loses followers very quickly. When Napoleon said, "CHARGE," it may have been his final word, but his followers believed in him.

- can conceptualize and follow a dream. An effective manager is one who establishes goals and then strives to close the gap between where they are now and where they want to be in the future. Closing this gap requires careful planning.

- has strong communication skills. An effective manager must be able to adapt language and communication styles to suit different audiences. This manager must also be able to effectively summarize discussions and bring a group to closure on issues.

Human Relations — A good manager:

- has "social power." That is, a good manager needs to be accepted as a member of the top management team. This can be difficult because this person has the ultimate authority.

- can manage the group process. This means that a good manager is someone who can instill loyalty and pride among personnel. The Japanese are particularly good at this.

- has positive regard. A good manager sees a little bit of good in all personnel and openly expresses this positive regard. Minute managers lose this positive regard. They have lost complete ability to help personnel develop themselves.

- has objectivity. Good managers must listen to all sides of an issue and maintain a balanced perspective, as well as be objective about themselves. Strong managers and leaders are highly conscious of their own strengths and weaknesses and their need for development.

- has self control. A manager needs to remain calm under even the most difficult situations and make people believe that they are in complete control of the situation. You wouldn't find Napoleon up on his white horse saying, "Shut off those guns so I can think."

- has stamina. An effective manager must be able to keep the creative energy flowing, despite long hours each day.

Supervision — A good manager:

- recognizes and accepts accountability. A manager must accept accountability – this is basic to management. Participation is a yardstick with 35 inches of listening, giving, exchanging, tolerating, correcting, and adjusting and only one lonesome inch of decision making. No organization is truly a democracy. It is more like an autocracy where a good manager uses as many democratic methods as are effective. The more managers launder off that one inch of authority, the more trouble they will find themselves in.

- develops personnel. A good manager doesn't motivate people – people have to motivate themselves. A good manager does create the climate of conditions where people can motivate themselves. This is probably one of the toughest challenges facing managers in the public sector. By always being accessible, all managers can help personnel develop themselves.

- can make tough decisions. A good manager must be able to make tough decisions – quickly if necessary. Moreover, a good manager must be able to say no even if it hurts for a little bit.

This information was summarized from the ICMA Annual Conference session, *James L. Hayes – Take Two*, held on Monday, October 28, 1985, in Philadelphia, Pennsylvania.

The difference between Managing and Leading

Manager	Leader
Cop	Cheerleader
Referee	Enthusiast
Devil's Advocate	Nurturer of Champions
Dispassionate Analyst	Hero Finder
Professional	Wanderer
Decision Maker	Dramatist
Naysayer	Coach
Pronouncer	Facilitator
	Builder

The Leader must have infectious optimism – the final test of a leader is the feeling people have when they leave the leader's presence after a session. Do they have a feeling of uplift and confidence?

The Leader...

...must have infectious optimism – the final test of a leader is the feeling people have when they leave the leader's presence after a session. Do they have a feeling of uplift and confidence?

Notes ▪

Developing And Maintaining Discipline

13

Obey Instructions

You have all probably heard one or more of the following statements:
- "The department lacks the necessary discipline."
- "The Chief wants every officer to be more strict on discipline."
- "The father should be the boss of the home."

We have all heard these statements over and over until the very word "discipline" makes us think of punishment.

But discipline means much more than this. The word itself comes from disciple – that is, one who follows the teachings and examples of a respected leader. This gives some clue as to the true meaning of discipline.

- A well-disciplined fire company is one that has learned the first lesson of supervision: Obey instructions.
- A well-disciplined military drill team has learned the importance of obedience to command. Well executed military drills, which amaze and delight onlookers, are the result of instructions repeated many times over.

As Fire Department Officers, let's see what discipline means to us.

The True Nature of Discipline

Discipline is:

1. Regular training for obedience or efficiency.

2. A system of planning, orderly control, and conduct.

3. Punishment or rewards.

Essentials for effective discipline

Effective discipline comes with good supervision. Each volunteer must know what they are supposed to do; how they are supposed to do it; and why they are supposed to do it.

Disciplinary rules should be made known to all volunteers.

These rules should state the conditions under which corrective actions will be taken, what they consist of, and the department and city or fire district position regarding fairness, consistency, adequate investigation and rights of appeal.

There must be a clear definition of the responsibilities of all personnel in disciplinary matters.

Special emphasis must be given to the disciplinary jobs of supervisors, top management, and the personnel department or the civil service commission.

Records

Complete records of corrective actions usually are needed.

Records should include personnel identification, nature of each offense, a description of the investigation, action taken, and results of the action.

Misconduct can usually be attributed to lack of planning on the part of supervisors. When activities are scheduled and planned in such a manner as to keep volunteers active, there is little or no time for misconduct.

Organize activities around a time schedule and stick to it. This includes drills, as well as special events. If a training program is scheduled to begin at 7:00 p.m., start it promptly at 7:00. People who value time appreciate others being prompt.

Major causes of misconduct

The major reasons for volunteer misconduct are:

- Boredom.
- Discontent.
- Idleness.
- Lack of interest in the job.
- Lack of work and assignments.
- Inadequate supervision.
- Misunderstanding of policies and their need and purpose.
- Lack of uniform enforcement of regulations.
- Resentment.
- Poor communication.
- Emotional strain.

The wise officer can avoid much formal discipline by doing all they can to remove such causes.

Positive or negative discipline

Positive or constructive discipline is the force that comes from inside and prompts the personnel to obey rules and regulations. People in this type of work group do what is right because they do not want to hurt the group.

Negative discipline, well known in the old days and still practiced by the driver-type supervisor, is the discipline of fear. It is based on threats of punishment.

The main drawback of this method is that a volunteer subjected to this kind of supervision will do only enough to get by when you are watching, or they will quit.

Discipline and morale go hand in hand.

Positive discipline is closely related to the admiration and respect of volunteers towards their supervisor.

Human relations

The relationship that exists between the officers and their personnel is a powerful force in building positive discipline.

The relationship is ideal where the officer appreciates and understands their people; where they have their interest and welfare at heart; where they respect their opinions, knowledge and skills. The following points will help you, as an officer, to understand and practice human relations for increased efficiency and better working relations with your people:

1. The officer must understand the principles, rules and regulations that promote good conduct and practice them.

2. The officer must know their volunteers as individuals be consistent with them and treat them fairly and impartially.

3. The officer must develop a sense of belonging in the group.

4. The officer gets information to their personnel promptly and accurately. They must work to eliminate rumors.

5. The officer uses their authority sparingly and without displaying it.

6. The officer must learn to delegate authority as far down the line as possible.

7. The officer seldom makes an issue of minor infractions, nor do they make personal issues out of discipline.

8. The officer displays confidence in their people, rather than suspicion. Their personnel will be faithful to an officer who has expressed confidence in them.

9. The officer must train their group well.

10. The officer has given attention to the mental and physical welfare of the group.

11. The officer must try to avoid errors, but they must also show willingness to admit when they make mistakes.

12. The officer must develop loyalty in the group.

13. The officer must realize that idleness leads to dissatisfaction. Keep volunteers busy doing important things. Their time is valuable.

14. The officer knows that discipline cannot be completely standardized due to individual differences. No two persons are alike in all ways.

The base for positive discipline

Foundation

Good discipline does not just happen. It takes good organization and supervision along with a well understood group of objectives to make a modern, alert and orderly organization.

Firefighters must take pride in their work. Lack of such pride will surely mean low-quality work and poor morale.

Atmosphere

Working conditions must be good. The working environment must be friendly. Officers must take an active interest in the mental and physical health of each person.

Attitudes

Dictatorial attitudes do not go with proper supervision. Every volunteer must feel free to discuss their problems and their complaints with their superior officer. The volunteer also needs an appeals procedure to assure justice and fair practice. Too many officers think that discipline means punishment only. There is a much better interpretation. Discipline is a connecting force, a team attitude. Discipline results from, and results in, high morale and cooperative efforts to achieve agreed-upon objectives.

Discipline problems are largely problems of training. All personnel must learn the reasons for necessary rules. Well-disciplined volunteers understand that rules are necessary for the success of their operations. There must be respect for the volunteer as a person. There must be respect for the group and belief in its ability to enforce its own discipline. Your challenge as an officer is to discover and eliminate the underlying causes of faulty behavior.

Responsibility

The major and final responsibility for sound and lasting discipline belongs to top management.

The Company Officer, the Department Head, and Staff Officer, however, must help in fulfilling this responsibility. All officers must establish the kind of working relationships that will make volunteers want to do their best work.

This does not mean that Officers should be too lenient.

Officers instead must be both management-minded and volunteer-oriented.

Guides to building positive discipline

- **Set a good example**
  Good volunteer behavior begins with good discipline, and volunteers expect officers to set a good example.

- **Let the volunteer know**
  Volunteers must know the expected standards of conduct and performance and why they are required.

- **Create a good work atmosphere**
  Volunteers perform their best work in a friendly atmosphere. All volunteers must be able to count on fair treatment. Their suggestions should be welcomed so that problems can be corrected. Be sure to recognize individual initiative and give credit for it.

- **Offer counseling**
  In handling problem volunteers, you should try to promote understanding and confidence. You should be specific in describing what you do not like about the volunteer's behavior or work. Let the volunteer give their side of the story and then help them decide how they can improve.
  The volunteer should feel that both of you have worked together to discover the problem and how to correct it. Then it should be clearly understood by both of you that specific, definite improvement must be shown within a fixed period of time.

- **Maintain firm and impartial control**
  Volunteers, as a group, should accept considerable responsibility for maintaining their own discipline. This will occur if they understand the rules, feel that they are reasonable and enforced with complete fairness.
  The uniform, impartial handling of all problems builds respect and prevents serious grievances from developing.
  Don't ever play favorites.

- **Take action when you have to**
  Regardless of how understanding and skillful you may be, it is sometimes necessary to take action.

Maybe you can only recommend that disciplinary action be taken, but you still have a definite and critical responsibility to make specific recommendations. You are the one who can best prescribe the disciplinary action after seeing the problem from a first-hand position, and can recommend action that will be most constructive for the volunteer.

Be sure you know what kinds of formal corrective actions may be taken and what your authority is. If you do not know, find out by talking with your superior officer. In most fire departments, the officer has the major responsibility for disciplinary recommendations. They must also provide information for management so that they can take action if they have to. Each officer, therefore, needs to keep accurate records of Dates, Times, Places, Circumstances, Witnesses, relating to violations of rules and standards of conduct.

Taking corrective action

Typical steps:

Even with constructive discipline, volunteer misconduct will occur, and corrective action becomes necessary.

The following kinds of actions show what you can do, starting with the mildest action and going through dismissal.

1. **Warning.** This is simply a matter of telling the volunteer that their behavior or performance must improve or more serious action will be taken. The action may be written or oral, formal or informal. In such cases, a specific, detailed record should be kept of each important warning given the volunteer. The warning should be in private and always should come before any of the other actions described below (unless serious offense is committed).

2. **Reprimand.** A reprimand is a formal record of an interview with a volunteer who has been told that more serious action will be taken unless there is immediate improvement in their performance or behavior.

   The reprimand usually should be issued by a higher level of supervision than the first-line officer. Usually, there should be a witness to the interview. A copy of the record of reprimand should be given to the volunteer. A sufficient number of copies of the reprimand should be prepared for all persons concerned. The report, in addition to proper identification of all parties, should include, where appropriate, the following items:

   a. The behavior or work performance for which the reprimand is given and the specific improvements expected.

   b. The time limit within which improvement must be made.

   c. A specific offer on the part of the officers and management to assist the volunteer in the expected improvement.

   d. A statement of any prior warnings given the volunteer.

   e. Some indication of further action to be taken for insufficient improvement.

3. **Suspension.** This is the most serious action that can be taken prior to discharge. An interview with the volunteer is recommended. Suspension should be subject to review by top officers or management.

Be sure you know what kinds of formal corrective actions may be taken and what your authority is. If you do not know, find out by talking with your superior officer.

4. **Demotion or Discharge.** Usually, the Fire Chief is the only authority who may discharge or demote a volunteer. In the Fire District, it may be the fire district commissioners. Action of this sort should be taken only as a last resort.

Practical helps to discipline

When an order is disobeyed, you are not doing your job unless you do something about it. Probably the most common type of discipline used by fire officers is a simple warning. In giving a warning, though, it must be fitted to the individual and the situation.

Just the slightest hint of something wrong will be more crushing to the sensitive individual than the tongue-lashing you might give the thick-skinned volunteer.

The warning should be calm, constructive action. It is not intended to be destructive. You are interested in building people, not tearing them down. Be interested in the underlying causes, not how to get even with anybody.

Failure to warn when it is due is also bad. The officer who is too soft is ineffective. If one volunteer gets by with something, you may lose control. Too many warnings are just as bad.

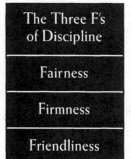

The Three F's
of Discipline

Fairness

Firmness

Friendliness

The three "F's" of discipline are a good guide. These are Fairness — Firmness — Friendliness.

The officer who follows these will have few problems.

Recommended procedure for the warning:

1. Get all the facts.

2. Warn in private – never in front of others.

3. Put the volunteer at ease. Give a word of praise first, if you can, to take out the sting.

4. Use no sarcasm, anger or abuse.

5. Fit the warning to the individual.

6. Lay out the facts.

7. Ask rationale behind performance.

8. Do not threaten the volunteer (they know just how far you can go).

9. Once they admit they were wrong, the warning is over.

10. Leave on a friendly note. Let them know the incident is closed. No Nagging!

11. Follow up later with a casual and friendly contact at the volunteer's work station, but do not mention the warning.

Ask yourself this question to determine the effectiveness of your warning: Did it correct the situation? Remember, you need to get along with this volunteer in the future. You must keep them working and producing.

You must be able to get along with your own conscience. There is a great difference between dignity and arrogance! "Discipline is the kind of supervision which makes punishment unnecessary."

Morale building

High morale is indicated by a feeling of satisfaction, pride of membership or by a general happiness.

This feeling is developed through the fulfillment of the natural desires of people.

People are born with natural desires, listed briefly as follows:

1. A desire for food.

2. A desire for social approval or recognition.

3. A desire for sex satisfaction.

4. A desire for a home and family.

5. A desire for security.

6. A desire for recognition and award.

Satisfy the above and you develop high morale.

Let us convert these desires to things volunteers want. They will list them as follows:

1. Recognition for a job well done.

2. Challenge.

3. Training.

4. A sense of accomplishment.

5. A sense of belonging.

6. A feeling of importance in the organization.

7. A knowledge of just how they stand.

8. An opportunity to advance.

9. Have fun!

Help them to get what they want, and you have high morale and a smoother operating machine that turns out increased production. You thereby get the things you want.

Notes ■

Coaching For Achievement

14

"Six Steps For Success"

Normally, "discipline" is regarded as a distasteful task which, if done, is done with reluctance. Sometimes, it may appear easier to simply "look the other way" when it comes to unacceptable behavior within your volunteer organization. While you may think, "What I don't know won't hurt me...and if I don't see it – I don't know it," unacceptable behavior seldom goes away. In actuality the failure to deal with unacceptable behavior reinforces it.

To compound the problem, we all too often fail to reinforce acceptable behavior. When someone does something right, it is expected. You may even have heard the comment, "I'll talk to them when...they screw up!" We must not take good people and their behavior for granted. Let them know that you appreciate them. When people become dissatisfied, and oftentimes angry, with their fire department, work place, friends or family, it just may be that they feel they are taken for granted – that they are not appreciated.

The irony to all of this is that reinforcing acceptable behavior is simple, easy, inexpensive, it "grows" people, the recipient feels good and the giver feels good.

So why don't we do it more often?

You cannot change what has gone on in the past. You can control what you choose to do in the future...starting with **Now**!

Coaching for achievement has two objectives:

1. To provide you, the person responsible for managing volunteer firefighters, emergency medical responders and support personnel, with communication tools to help your volunteers perform and be successful.

2. To assist you in being successful in your role as an officer. Coaching for Achievement has its foundation on five guiding principles:

Principle No. 1: The Fire Officer is a coach

As a Fire Officer, you have a leadership role in your organization. Included in that leadership role is the building of winning teams and the development of people. In developing individuals, you need to provide a climate in which the volunteer can grow, recognize and meet the needs of the individual, reinforce good behavior, correct unacceptable behavior, and set an example for others to follow.

Principle No. 2: Volunteers are valuable

If you and your organization have taken the time and money to carefully recruit, select, and train a volunteer, then you have a sizable investment in that person. Also, people should be offered the opportunity to be successful. Further, each person has been endowed with unique talents and goodness. (Granted, some people may be hiding those treasures; nonetheless, it is true.)

Given this valuable volunteer, you do not want to lose them. Therefore, you are willing to work through a process which benefits both the organization, you the officer, and the volunteer member.

"Good Leaders never walk by a mistake".

General H. Norman Schwarzkopf, Jr.

You cannot change what has gone on in the past. You can control what you choose to do in the future... starting with Now!

If for some valid reason you find that this volunteer is no longer a valuable member who contributes to the success of the organization (and that both they and you would be happier if they were somewhere else other than your organization), then do not make any further investments in coaching this person. Terminate the individual from your organization before you, they, and your fire department suffer any further grief.

Principle No. 3: The "Spice Theory" of discipline

Spice, by itself, is bitter and hard to swallow. A small amount goes a long way. However, if it is properly mixed with other good ingredients, a wonderful result can be produced.

Negative discipline, by itself, is bitter and hard to swallow. A small amount goes a long way. However, if it is properly mixed with positive discipline, a wonderful result can be produced.

Principle No. 4: The "Milk of Human Kindness"

A relationship between two individuals is likened to a glass container. Into that glass, both parties pour the "milk of human kindness." Only when the glass container is equally filled to top by both persons can the relationship be full, honest and productive. The failure of either person to contribute fully to the relationship prevents it from reaching its potential.

The glass container can spring a leak, spill its contents, become cracked or even broken. It can be repaired, sometimes with great difficulty. It can be cared for or discarded.

Principle No. 5: The responsibility for our choices

We are presented with situations every day requiring us to make a choice (even a choice of doing nothing). Each choice we make results in consequences. We are responsible for the choices we make – good or bad!

- We are responsible for choosing the example to set for others to follow.

- We are responsible for how we choose to look at others and see their talents (or defects) and the good (or bad) which they possess.

- We are responsible for how we choose to administer discipline (positive and negative).

- We are responsible for the choices we make in developing relationships with others.

We are responsible. Even though there are times we would like to blame someone else. Of course, that is our choice, too. Isn't it?

There are six steps to success in coaching for achievements:

1. Identify the **behavior**

2. Decide if the behavior is **acceptable** or **unacceptable**

3. Evaluate the possible **reason** for the behavior

4. **Confront** the unacceptable behavior

5. Reach an **agreement**

6. **Monitor** the volunteer's behavior and provide feedback

Negative discipline, by itself, is bitter and hard to swallow. A small amount goes a long way. However, if it is properly mixed with positive discipline, a wonderful result can be produced.

Step One — Identify the behavior

Before you can discuss the behavior with the volunteer, you must first identify what it is that the member is doing. You should be able to write down the specific behavior or action on one side of a 3" x 5" card.

Most important, throughout this process, especially when you deal with unacceptable behavior, you must ask:

- Is your source of information relating to the action or incident?
- Do you have personal knowledge of the volunteer's behavior?
- Do you have witnesses to the volunteer's behavior?
- How accurate and impartial is information provided?
- Are they providing something they observed or are they giving you an opinion?
- If it is serious enough behavior, is the witness willing to give you a written statement?
- Do you have physical evidence that connects the volunteer to the behavior or consequence of the behavior?
- Do you have inferences that can be logically drawn from known factual information which points to the volunteer?
- Do you have additional information from more than one source which strengthens and reinforces your conclusion regarding the volunteer's behavior?

Step Two — Decide if the behavior is acceptable or unacceptable

- Why is the behavior acceptable or unacceptable?
- What standard or criteria are you measuring the performance against?
- Is it a fair, impartial and objective evaluation of the behavior?
- Make certain that you communicate to all of your volunteers what is expected.

If the behavior is determined acceptable, it should not be taken for granted. All too often, when we see acceptable behavior, we fail to reinforce it. Just as unacceptable behavior needs to be corrected, acceptable behavior needs to be reinforced and rewarded. By supporting acceptable behavior, we send a message, not only to that individual volunteer, but throughout the organization of what is expected and appreciated.

The same six steps apply to acceptable behavior. They are much more fun to deal with however:

step 1. Identify the behavior.

step 2. Determine why the behavior is acceptable.

step 3. Evaluate the reason for the behavior. (Learn from it and repeat it as often as necessary.)

step 4. Confront the behavior. (Tell the volunteer member that what they did is appreciated.)

step 5. Reach an agreement. (Reaffirm that the behavior is expected to be repeated in the future.)

step 6. Monitor and provide feedback. (Continue to reinforce and praise acceptable behavior. Don't take your volunteers for granted; let them know you appreciate them and their efforts.)

If the behavior is determined unacceptable, there should be a criteria that identifies why it is unacceptable behavior within your volunteer organization.

In making your determination and reaching a conclusion you will need to determine if the unacceptable behavior was:

- the result of circumstances beyond the control of the volunteer member

- the result of an unintentional act

- the result of negligence (or carelessness)

- intentional and possibly malicious

Step Three — Evaluate the possible reason for the behavior

This is important for unacceptable behavior. If the behavior is unacceptable:

- First, determine if the person has the necessary physical, mental and emotional ability to function successfully in the assignment they have been given.

- Second, determine if the person has the necessary instruction and training to successfully perform.

- Third, if the volunteer has the ability and the training to behave in an acceptable manner, it would appear, most likely, that the individual is choosing to behave in a manner that is unacceptable.

A volunteer can choose to behave in an unacceptable manner because they are dissatisfied. They can be dissatisfied for a variety of reasons:

- the responsibilities assigned to and expectations of the volunteer are unclear or non-specific

- the required completion of unnecessary or confusing paperwork

- burdensome bureaucratic procedures, sometimes known as "red tape"

- unfavorable working environment, broken or defective equipment or the physical environment

- a new officer (supervisor) or transfer to another fire company

- an incompetent or dishonest officer (supervisor)

- frustration because of an inability to accomplish objectives which are out of reach

- boredom because of a lack of challenge

- apathy because of a lack of focus, mission or purpose

- giving the volunteer an inappropriate or just plain wrong assignment

- a physical disorder

- emotional trauma or distress

- a perception that they are not appreciated

Occasionally, the question is raised about "bizarre" behavior. Specifically, where do you draw the line in that gray area of having fun – practical joking – destructive action? The following guidelines can help in answering the question:

- if it damages equipment
- if it damages the public image of the fire department
- if it damages the morale of the volunteers
- if it disrupts the working environment
- if it creates a hazard or safety problem
- if the person knows better
- if it violates a law

If any of the above are present, then the behavior has crossed over the line. It should be treated as an unacceptable behavior and dealt with in an appropriate manner.

Step Four — Confront the unacceptable behavior

In step 4, you meet with the volunteer and confront the behavior that is unacceptable. At this point, you are not in an adversarial role. You may be later, but not now! The purpose of this meeting is to:

- communicate with the volunteer
- develop a solution
- reach an agreement which is acceptable to both parties
- reaffirm to the volunteer that they are a valuable member of the team and needed for the success of the fire department

Unacceptable behavior is not acceptable because it prevents them and the group from being productive.

Where do you meet?

The meeting place needs to be quiet, where you will not be interrupted.

The meeting place needs to be non-threatening, where the volunteer will not feel defensive nor intimidated. Remember, the purpose of the coaching session is to communicate and correct unacceptable behavior.

The meeting place should be comfortable. Avoid having a barrier, such as a desk, between you and the volunteer. Sometimes, a chalkboard is beneficial. Both you and the volunteer can use the board. Also, both of you are looking at the board together rather than opposing each other.

How do you start the meeting?

Show a mutual respect to the individual when they enter the room. Stand up to greet them, shake their hand and thank them for coming to the meeting. Ask them to sit down.

This may sound very basic. It is very important that you set the proper mood. Your demeanor communicates more than what you say. Your sincerity and genuine interest in the volunteer will speak louder than your words.

It is important that you do not make accusatory statements. Possible opening lines which can be used include:

- "There is a situation that I need your help with."

- "I need your assistance in solving a problem."

- "I'm concerned about an incident that occurred."

- "I have some information and I would like to know if it is correct or not."

Remember, you are attempting to solve the problem, not increase the tension.

How do you describe the problem?

Describe the behavior in a factual and chronological manner. Approach the problem first in the past, then the present, then the future.

Focusing on the specific behavior or action, describe what occurred and the result or consequence.

- Ask the volunteer if the information you have is correct and accurate.

- Ask them if they have information to add that might help you understand what happened.

- You might ask what caused them to act in such a manner.

Next, advise them on how you feel about their behavior and why you are concerned. Make it clear to them you are upset with their behavior. Make it clear they are a valuable member of the team and this is not conductive to the good of the organization.

It is important to **Listen** to the volunteer. That individual, if a valuable member of your team, deserves a fair hearing. As you sit quietly and listen, you may need to periodically summarize what the person has said. Phrases which can help you in successfully doing that include:

- "From your point of view..."

- "As you see it..."

- "You think..."

- "You feel..."

- "What I hear you saying is..."

- "Could it be..."

- "It appears..."

- "Let me see if I understand you..."

Now, explain to them your expectations of them and their behavior in the future. That instead of dwelling on the past.

Step Five — Reach an agreement

Make certain that both you and the volunteer understand what has been agreed to for behavior in the future. Depending upon the behavior, the circumstances and the policy of your fire department, there may be no need for discussion of a solution.

If the future expectations are not clearly identified, it is best to involve the volunteer in developing the "terms" of your agreement. It should be clear what the "terms" are and the roles that both you and the volunteer will play in honoring the agreement.

It is expected that this coaching session and agreement will resolve the problem and never be repeated. However, it would not be totally honest unless all the cards are put on the table. An important part of your agreement is the understanding of what will happen if the unacceptable behavior is repeated in the future.

The importance of specifying a consequence for future unacceptable behavior or actions is not only making your position clear but also clarifying your role in this process. Instead of the burden being on you to make a fair decision on future problems (and, in effect, being the judge, jury and executioner), you are honoring the agreement that both you and the volunteer made prior to the repeat of the unacceptable behavior. There are no surprises nor hidden agendas.

You should finalize your agreement with a hand shake, a written memorandum or a written letter of agreement that both you and the volunteer sign. The extent of formalizing the agreement is dependent upon the trust which exists between you and the volunteer.

A written agreement should factually include:

• the date of the coaching session.

• the unacceptable behavior and results of the behavior.

• your concern of the unacceptable behavior.

• future expectations of acceptable behavior.

• consequences, if the unacceptable behavior is repeated.

• a provision that if the volunteer fulfills their part of the agreement (no repeat of unacceptable behavior or actions) within a specific time period (30 days, 60 days, 90 days, etc.) the written memorandum or letter of agreement will be removed from their personnel file.

Step Six — Monitor the volunteer's behavior and provide feedback

It is important that you follow up on how the person is doing. After all, you and the volunteer have an agreement. If the person is being successful in honoring their part of the agreement, you need to congratulate them. If there is a problem of recurring unacceptable behavior, you must honor your part of the contract by enforcing the agreement.

When providing feedback, there are five important parts:

1. **Time** — provide feedback as soon as possible after the occurrence. Keep in mind, however, the location and surrounding events may cause you to delay the feedback until a more appropriate place and time.

2. **Place** — "praise in public, reprimand in private" is always sound advice.

3. **Behavior** — speak to specific behaviors and actions which the volunteer has control over and can change.

4. **Amount** — provide feedback in bite-size pieces, one at a time, to the volunteer.

5. **Follow up** — carry out your part of the agreement. Do not allow the member to become "out of sight, out of mind." If there is a reoccurrence, although there might be an urge to grab them by the throat and tell them, in no uncertain terms, how upset you are with them, you must refrain from doing so.

Remind the volunteer of the agreement you both have. Ask them if there has been something extraordinary which altered the agreement. If there has not been (which is most likely the case), you need to inform the volunteer the agreement was important to you and you will honor your part in the agreement – enforcing the consequences for repeating the unacceptable behavior. If the unacceptable behavior or action is further repeated, appropriate disciplinary action is necessary. You may want to offer the volunteer an opportunity to make a critical choice: either choose to behave in an acceptable manner or choose to become a former member of the fire department. These types of critical decisions should always be in writing and signed by all affected parties.

Summary – Coaching for achievement

Principle No. 1

The Fire Officer is a coach

You have a responsibility to build and develop your volunteers. As the coach you build a winning team by building on the strengths of the volunteer members.

Principle No. 2

Volunteers are valuable

Given the time and money that has been expended in the recruitment, selection and training, you have a major investment. Your volunteers are your most valuable resource.

Principle No. 3

The "Spice Theory" of discipline

A small amount goes a long way. Negative (punitive) discipline needs to be blended in with large amounts of positive (recognition and reinforcement) discipline to produce the desired results.

Principle No. 4

The "Milk of Human Kindness"

Both individuals must equally contribute and fully fill the glass container with the "milk of human kindness" to have an honest and trusting relationship.

Principle No. 5

The Responsibility for our choices

You are faced with choices everyday. Whatever the choice you make, you are accountable for your choice and your behavior.

There are six steps to success in coaching for achievement. When the behavior is acceptable:

step 1. Identify the behavior.

step 2. Determine why the behavior is acceptable.

step 3. Evaluate the reason for the behavior.
   • Learn from it and repeat it as often as necessary.

step 4. Confront the behavior.
   • Tell the volunteer member that what they did is appreciated.

step 5. Reach an agreement.
   • Reaffirm that the behavior is expected to be repeated in the future.

You have a responsibility to build and develop your volunteers. As the coach you build a winning team by building on the strengths of the volunteer members.

step 6. Monitor and provide feedback.
- Continue to reinforce and praise acceptable behavior – Don't take your volunteers for granted; let them know you appreciate them and their efforts.

When the behavior is unacceptable:

step 1. Identify the behavior.
- Do you have the facts – all the facts?

step 2. Decide if the behavior is acceptable/unacceptable.
- What makes the behavior unacceptable?
- Is it a judgment that is fair, impartial and objective?

step 3. Evaluate the possible reason for the behavior.
- Are you dealing with the cause or merely the symptom of the problem behavior?

step 4. Confront the unacceptable behavior.
- Confront the unacceptable behavior by meeting and communicating with the volunteer.

step 5. Reach an agreement.
- Cooperatively solve the problem of the unacceptable behavior. Include in your agreement the specific consequences for repeating the unacceptable behavior in the future.

step 6. Monitor the volunteer's behavior and provide feedback.
- Help the volunteer member to live up to the expectations of the organization and be successful.

Notes ▬

Team Building

15

A Ten Point Program

A man once bought a team of four strong horses. He then acquired a handcrafted wagon constructed of the finest materials. He purchased expensive harness and rigging for his team. He then proceeded to outfit the team and hitch them to his wagon.

He hitched two of the horses to one end of the wagon facing northbound and the other two horses to the opposite end of the wagon facing southbound.

The gentleman then climbed up into the wagon where he proceeded to scream and shout while cracking the whip. The horses sweated and strained and caused a cloud of dust while the wagon violently shook, rattled to and fro, and rolled back and forth. Climbing off the wagon, the man proclaimed in disgust, "This outfit isn't worth a damn."

The story sounds ridiculous. Anyone watching would have told the man that his horses were not pulling in the same direction. After all, everyone knows that a team of horses moves in the same direction. It is not too difficult to gather a number of individuals together. It may not be that difficult to get them to go through the motions of appearing to be a group. It takes a lot of work to shape those individuals into a team. It takes even more to make them into a winning team.

There are ten basic points to be covered in starting to build your winning team:

1. Clarify the values of your fire department

What do your volunteers and organization stand for? What values do you believe in? The structure, programs and operational plans of your volunteer organization are built upon a foundation. The solidarity and strength of the foundation are a measure of the values which the membership, as a whole, believes to be important.

Examples of the values which winning teams believe in are:

- Pride in performance

- Loyalty to the success of the organization

- Teamwork through cooperation with other members

- Self discipline and accountability for behavior

- Dedication through a willingness to work hard

- Trust through building honest relationships with other members

- Credibility by being consistently accurate and honest

- Dependability by being there when needed, especially in the tough times

Without a belief in the values, it is doubtful the volunteer will work hard to make the goals and objectives become a successful reality.

2. Clarify the purpose of the fire department

What are the reasons that your fire department exists? A traditional starting point is "to protect life and property from the destruction of fire." But, as we all well know, modern fire departments are doing that and more. The purpose can be viewed from three perspectives:

a. Legally mandated responsibilities

As an organization, what services and functions are you legally obligated to provide? Those obligations include mandates by state law and local ordinances.

b. The needs of the community

Factors influencing the needs of the community include the population of the area, the age distribution of the population, population density and distribution, high risk occupancies, transportation corridors, geographic features, climactic conditions and economic growth.

c. The expectations of the people

- What are the expectations of the community?

- What are the expectations of the elected officials?

- What are the expectations of the business community and industry within your response area?

- Is there an expectation for advanced levels of emergency medical services, specialized rescue services, and hazardous materials response capabilities?

3. Clearly identify the mission of the fire department

The mission expands upon the purpose. It identifies the role of your Volunteer Fire Department in serving your community. The statement of the mission should be clear and concise. A mission statement says who you are and what you are about. The following is an example of a mission statement:

"Your local (name) Volunteer Fire Department is a professional and progressive organization, (number) active members strong, who are committed to serving the people of our community by providing rapid response, effective operations and quality care in the event of fire, rescue and medical emergencies occurring in our area." In writing a mission statement, try and keep it short and simple.

Accompanying the mission statement should be a slogan which reinforces the mission statement of your organization. The "Elk Station Fire Company" in Spokane, Washington, Fire District No. 4 has the following for their slogan:

"The Desire to Serve

The Ability to Perform and

The Courage to Act"

4. Set goals for your fire department

Goals provide a specific direction for the organization to move in, and establish limits on the range of activities to be undertaken by the fire department or one of its group.

Goals are stated in broad general terms. A goal statement, by being open-ended, provides continuity to the program. Consequently, it does not have a specific date for completion. A goal statement begins with such verbs as "support," "maintain," "assist," "provide," "monitor." Examples of goal statements are:

- "Provide quality emergency medical services."
- "Present the best possible image to our customer, the public."
- "Provide hazardous material emergency response to the industrial area."

The goal statements should be posted in several locations for all members and the public to read.

5. Establish priorities for your fire department

What areas of performance are critical to the success of your volunteer fire department? If your fire department had to stop doing something tomorrow, what would it be? What next would be stopped or suspended?

Sadly, there are times when people cannot tell you what their priorities are. They do something here, then something there, but never seem to complete the program or project. Even when someone expends a lot of effort, they just can't seem to accomplish the desired results.

There are times when you simply cannot do everything you desire to do. When faced with limited resources, a lack of time, or both, you identify the most important task, assign it a priority, throw your resources at it and get the job done.

You need to establish priorities before you reach a time of crisis. This is an important part of the planning process. Priorities are especially important when budgeting for your department's operation. If the limitations on funding require a reduction, you start at the bottom of your list and work upward until the revenue and expenditures balance. If the priorities are decided upon first, it avoids a last minute mad scramble to decide what to eliminate from the budget.

In establishing priorities, you are making a commitment. You are saying that you will focus your resources for maximum effectiveness. Further, you will not dilute their effectiveness by spreading them too thin and, in effect, wasting them on areas that are not important to the success of your operation.

Your fire department's priorities should be reduced to written form (listing them in their order of importance) and posted for all members to view.

6. Set operational objectives for your fire department to achieve

Operational objectives are targets to aim at and shoot for. Objectives, which are developed from a goal statement, specify the results to be achieved. They answer the questions of "What is to be accomplished?" and "When is it to be accomplished?"

It is important that objectives be realistic and attainable. While goals are general, objectives are specific. While goals are open ended, objectives can be verified and measured. Objectives specify quantity, quality and time. Objectives are stated in end results and can begin with verbs such as "create," "plan," "develop," "establish," "implement" and "design."

Operational objectives provide the action plan for the team. As such, the team members should be involved heavily in developing the objective statements. They are a tremendous source of information. And, if they are involved in the planning and agree to the end results, it is now their plan. The success of the plan becomes their personal success.

There are times when you simply cannot do everything you desire to do. When faced with limited resources, a lack of time, or both, you identify the most important task, assign it a priority, throw your resources at it and get the job done.

Examples of objectives for your recruiting program could include:

"Create and produce, by February 2, 20_, a ten (10) minute slide/tape presentation which explains fire department operations to persons expressing interest in becoming volunteer members."

"Present a two hour training session on the 20_ Recruitment Campaign to 90% of all fire departments members by March 1, 20_."

"Identify a minimum of 50 individuals who are interested in becoming a volunteer firefighter by May 15, 20_."

7. Communicate to each volunteer their assignment

Winning athletic teams have plays in which each member has an assignment to carry out. For the team to be successful, each member must effectively do their job. If a player cannot perform, they are replaced by someone who can. This is done in the best interest of the team as a whole.

Keep in mind, the team member must first know what their assignment is if they are to carry it out.

Are you communicating to each member of your team what their assignment is? Do both you and your team members understand and agree on:

a. their place within the organization?

b. what they are held accountable for, any authority they have to get their job done, the resources they have to do the job?

c. the standard of performance you expect from them?

The Operational Objectives specify the What and When. This stage of the team building specifies the Who, Where, How and Why. When explaining to each member their particular assignment, it is advantageous to have a detailed activity plan which shows how their part fits into the whole.

The combination of all these items provides a basis for evaluating the performance of the fire department, various groups and individual volunteers. It is a management tool for giving meaningful feedback to your team.

8. Coach for achievement

Coaching for Achievement has "six steps to success." (See Chapter 14, "Coaching for Achievement") When you coach your volunteers you recognize, reward, and reinforce acceptable behavior. You also recognize, confront and correct unacceptable behavior. It is a combination of communication, counseling, discipline and motivation skills.

9. Provide status reports on the program

To measure the progress toward the achievement of an objective, periodic checkpoints or "mileposts" are necessary. The "mileposts" let you know how you're doing. They are standards to measure against to determine if you are on schedule, ahead of schedule or need to "pick it up some."

Three key items in status reporting are:

• Who is going to be responsible for monitoring the program/project?

• How often are the reports going to be made?

• Where do the reports go?

Status reports provide valuable feedback on the success of the current efforts which can reinforce the activities. This causes a "springboard" effect to occur and gives momentum for the program or project. In addition, they provide the opportunity for minor corrections to be made before major problems develop.

10. Celebrate the successes

Celebrating the achievement of objectives gives you the opportunity to recognize contributions your volunteers have made to the success of the fire department. In doing so you reinforce the positive behavior of your members. Further, their work enables you to reward them for a job well done. Most of all, you have the enjoyment of telling them how much they and their efforts are appreciated.

Summary

- Providing Status Reports allows you to monitor and measure the progress and success of activities toward achieving the specific operational objective.

- Objectives give the volunteers a specific target to achieve.

- Priorities tell everyone what is important and not so important.

- Goals give direction for the fire department to move in.

- The Mission expands on the purpose of the fire department and gives clarity to its role in the community.

- The Purpose of the organization is its reason for existing.

- The Values, which the membership believes in, are the force which drives the volunteers to carry out the mission and achieve the objectives.

- Communicating, Coaching and Celebrating reinforces positive behavior, corrects unacceptable behavior, recognizes achievements and lets people know that they are appreciated.

Notes

Managerial Dimensions

16

Strengthen Your Ability

Each dimension of your management ability is important to your success and the success of your organization.

Strengthen your ability as a **Strategist** by looking to achievements in the future and having a vision of what your fire department can become.

Strengthen your ability as an **Administrator** by selecting the best people for different assignments. Monitor their activities and pay attention to details. Evaluate the progress of your organization on a regular basis.

Strengthen your ability as a **Problem Solver** by systematically identifying the cause of the program, generating alternative solutions and selecting the best alternative based on well thought out criteria.

Strengthen your ability as a **Facilitator** by innovating better ways of doing things, working with others to overcome their resistance to change and mediating differences between volunteer members.

Strengthen your ability as a **Coach** by placing a high priority on training, develop competence and confidence in your volunteers and make an organizational commitment to excellence through a team effort.

The Manager as a Strategist

The Strategist has a vision of what the organization can become.

Your effectiveness as a strategist depends upon your ability to develop an effective plan. A plan to take your organization into the future. Given the values that are important to the group and a clear mission, what are your volunteers shooting for? Goals give your organization and its members a direction to move and a target to hit. As a strategist, you develop annual operating goals. You should also be making long range plans.

• Where will your volunteers be three years from now?

• Where will your organization be five years from now?

• What needs to be accomplished?

Important to the development of the plan is the priority assigned to each goal and objective. What is the most important thing which needs to be accomplished? What is next?

By establishing

1. a clear direction,

2. targets to achieve,

3. definite priorities

your volunteers can focus on the job they need to get done.

The Manager as an Administrator

Like the conductor of an orchestra, the administrator makes sure the members play their parts, in unison, and play out the plan of the symphony.

In order for the plan to be achieved, it is the role of the Administrator to

1. organize the individual volunteer,

2. assign them roles and responsibilities, and

3. provide them with the necessary resources they need.

It is important the administrator matches the right person with the right assignment.

It is necessary for the administrator to monitor the activities of the different members, or groups, and receive progress reports from key individuals. Most important is the ability to evaluate the progress of the operation and make the proper adjustments, as needed, to ensure that everything meshes together.

The Manager as a Problem-Solver

Before you can solve a problem, you must know what it is. This means going beyond the symptoms of the problem and finding the root cause. Sometimes you will see someone "solve" a problem by putting a band-aid on it. It may appear better, on the surface, but has the infection been properly cared for? The symptoms are only indications of the real problem. Only by carefully and systematically investigating the problem and identifying the cause can you hope to effectively deal with it.

Once the cause of the problem has been identified, you are prepared to generate possible solutions to the problem. But what is the best solution? What criteria are you using to evaluate the alternatives? Who have you sought out for advice? Have you gathered individuals together for a brainstorming session?

Solving one problem may create another. Have you carefully considered the consequences of the solution you have decided upon? When you have logically selected the most favorable alternative, does it "feel" right? Intuitive judgment has an important role to play in decision-making.

Can you live with the decision that you have made? It is important in making your decision that it is consistent with your values and ethics. After all, you are the one who must ultimately live with your choice.

The Manager as a Facilitator

To be successful as a manager, you must be able to get things done. Sometimes even the best laid plans go up in smoke. When that happens, it is necessary to modify your course of action. It may be necessary to innovate. Can you facilitate an innovative approach to achieving an objective?

Another important part of facilitating is to bring about change. To remain strong, you and your organization must be progressive. If you fail to keep up, you and your volunteers will be left behind. And yet, people can be resistant to change. How well do you bring about change within your department and with your volunteers? Two keys in bringing about change are to

1. solicit input from the people affected, and

2. prepare them, in advance, for the change.

Finally, as a facilitator are you able to mediate conflicts between volunteers with different perspectives? Do you work well with both individuals and groups? Your role in this area can be extremely important in order to achieve harmony within the department. An effective technique you can use is "listen and list."

Listen to the volunteers and hear their concerns. Then list out the points of agreement and the points of difference. It is best to list them out on a chalk board where everyone can see them. Then discuss how each side can move to an acceptable position where they can agree or at least accept the differences.

The Manager as a Coach

The most important resource you have is your volunteers. As the manager of those resources, you want them to perform to their potential. As the manager, you must coach them. Coaching means three things:

1. increase their competence

2. develop their confidence in themselves

3. build them into a team

Increasing their competence means developing their technical skills and ability to perform the role and responsibilities assigned to them as a volunteer. A comprehensive training program which develops their skills and maintains their prescribed skill level is a high priority.

While a person may possess the knowledge and training to perform, they can lack the belief, in themselves, that they can successfully perform when the time comes. The training program for your volunteers must develop their self-confidence. Through repetition, simulation and actual experience they demonstrate, not only to you, but to themselves, that they can perform their assignment with proficiency.

As the Coach of the Team, you must bring together the individuals and shape them into a group which shares a common mission, respects the abilities of each team member and trusts in each other to carry out their role effectively.

The Manager as a Communicator

We communicate through our words, actions and deeds. What is the message that you are sending to others? Do you hear what your volunteers are really trying to tell you? How well do you listen? Some people believe that good communication is how well they speak and the size of the words they use. Good communication is clear and simple communication. And most importantly, it is how well you listen.

We communicate to our volunteers by the example that we set. Is your example the one you want them to follow? It is easy to talk values and ethics. It takes courage and integrity to demonstrate them. If you expect people to trust you, you must be willing to trust others. To develop loyalty in others, you must be loyal.

It is important to communicate your expectations to the members of your department. It is important to communicate your vision of where the fire department is headed. It is important to communicate to your volunteers how they are performing. It is important to communicate to them that their efforts are appreciated.

Notes ■

17

Public Relations

A Key To Your Department's Success

Private industry long ago recognized that public relations is a responsibility second only to production.

Each volunteer is made aware that it is their personal responsibility to promote good public relations.

In the past, government has been charged with lagging behind private businesses in public relations. Today, public relations in the government service have taken on a new meaning.

The way a police officer. makes an arrest, the way fire fighters go about assisting after an emergency, the way the counter clerk treats a visitor, the way the telephone receptionist handles a complaining citizen – these methods of behavior, and many other things that people do, contribute to the total public relations of the organization.

It is also in the cleanliness of the streets, the availability of parks and playgrounds, and the pick-up of trash and garbage.

Public relations include everything that firefighters do or fail to do which has an effect on public attitude and the public's impression of the fire service.

Today, most city and fire district personnel are very much aware of the importance of good public relations. Public opinion has a lot to do with our efforts to obtain financial support for the fire department.

Volunteers must be on the alert to set an example in good public relations themselves, and to see that others practice good public relations both on and off the emergency scene.

What is Meant by Public Relations

"Public Relations" includes everything that a city or fire district does or does not do which has any effect on public attitude. This can include:

- The kind and quality of both physical and personal services rendered.

- The publicity given to activities.

- The appearance of buildings, grounds, apparatus and equipment.

- The appearance, attitude, and actions of firefighters (on and off the job).

- The fairness, honesty and consistency of providing regulatory and law enforcement services.

The most important person: The citizen

A citizen is the most important person to come in contact with us in the fire service, in person, by mail or by the telephone.

A citizen does not cause an interruption of our work. They are the purpose of it. We are not doing them a favor by serving their needs. They are doing us a favor by giving us the opportunity to do so.

A citizen is not an outsider to our activities. They are part of it and the reason for it.

A citizen is not a cold statistic. They are a flesh and blood human being with feelings and emotions like our own, and they have biases and prejudices too!

A citizen is not someone to argue or match wits with. Nobody ever won an argument with a citizen!

A citizen is a person who brings in their wants. It is our job to handle them satisfactorily.

Volunteer — citizen contacts

It is not until the citizen meets a fire department through its personnel or its services that the department becomes real to them.

Often this meeting is through contact while securing fire permits, asking for information on hazardous situations or in some other way such as registering a complaint or seeking our service in time of emergency.

It is at these points of meeting that most people get acquainted with their fire department, form an impression of what it is, and react favorably or unfavorably toward it.

In every member-citizen contact, the officer should see to it that the following general principles are followed:

Show genuine interest

Firefighters must show a genuine interest in the citizen's problems.

Whether the contact is personal or over the telephone—Pay attention! Ask questions to show interest and to get the facts (but do not cross examine). Be open minded and show a friendly helpfulness.

Furnish quality information

Information should be clear, complete, and accurate.

Do not "pass the buck." One thing that will irritate a citizen more than anything else is to be sent from one office to another to get information, only to find many times that they are still in the wrong place.

Directing a citizen from one buck-passing official to another over the phone is just as bad. Learn the functions of your organization's operation so that you may direct or advise the citizen correctly. If you cannot answer a citizen's question, you should offer to find the answer and call back.

If you are asked a question in person and do not know the answer, find the answer or direct the citizen to the person who can answer the question.

Pronounce words clearly

Personnel who speak frequently to the public should be trained to speak clearly. Their words should be chosen carefully so as to avoid any offense. Conversation should be businesslike, friendly, and efficient. Voices should have a "helpful" rather than an "official" tone.

Politeness and courtesy count

It is important that politeness and courtesy be observed. This should be a genuine, friendly politeness. Officers should train their personnel in the rules of business etiquette. Then, see to it that they practice the rules with all citizens.

A citizen is not someone to argue or match wits with. Nobody ever won an argument with a citizen!

Appearance is important

For those who come in contact with the public, appearance is extremely important. It is very important that everyone be neat, orderly and efficient at all times.

Casual contacts

You and all other officers need to set a good example by training your volunteers in the casual contact with neighbors, relatives, storekeepers, church members, local organizations, and so on. People will classify the fire department and city or fire district by the few people they know who work for the department.

The person who criticizes their department, who gripes about conditions, who spreads stories of waste and inefficiency, can do much harm to the organization.

The volunteer who is always disorderly, dirty or sloppy discredits the whole organization. So does the one who never pays their debts or who otherwise violates the rules of good conduct.

A few people who have violated the rules of good public relations bring discredit upon other loyal, hard working volunteers. While officers and supervisors cannot police them all of the time, they can do a lot in helping correct the situation by setting a good example themselves.

Telephone courtesy

Every person who will ever answer a telephone on the job should be trained in telephone courtesy. Never assume that they know how to use the telephone.

Officers must see that volunteers answer the telephone correctly. Even though many of the calls on your phone may come from within the department, you never know who is on the other end.

Here are some actual ways in which untrained individuals have been observed to handle incoming phone calls. What would you think if you were a businessman placing a call to a fire department office?

- "This is Joe speaking." "Hello." (Does not identify organization or the person speaking to the outsider on the line.)
- As they lift the phone, the sound of loud laughter greets the citizen. Apparently someone in the office has told a joke. The person laughs out a "Hello." (Citizen wonders if the stories they have heard about public agencies and fire departments are true.)
- A person answers in a sloppy drawl, mumbles, maybe smoking, chewing gum. (Indicates inefficiency and sloppiness.)

How should the telephone be answered?

Always answer the telephone by giving your organization unit and name (also title, if appropriate). A pleasant "Good Morning" touch is also appropriate. You should decide on the variation according to where you are and what your unit assignment is. Instruct your people on what to respond with.

Examples: "Good morning – Fire Department," or "Inspection Division, Inspector Crawford speaking," or "Good morning, Station Two."

Listen carefully to the incoming conversation. Answer questions carefully and clearly in a businesslike voice. Let your tone of voice indicate a pleasant and anxious-to-help attitude.

Do not make the party wait endlessly while you look up something. Ask them if they would prefer to have you call back. If the person called is not in, offer to take a message, in order that the person may return the call. Get the caller's name, organization, telephone number, extension and address if appropriate. If someone phones while you are out, be sure to return the call promptly.

Always keep a pad and pencil by the phone to record information. Do not depend on your memory! Always make a note if you commit yourself to call someone back. End the conversation with a courteous "Thank you" or "Good-bye. "

Wait for the other party to hang up, then gently replace your phone. Never slam it down.

Business etiquette

The following is a list of rules of good business etiquette which volunteers should follow.

Introductions

When being introduced to someone, or when introducing yourself, extend your hand in a firm, warm handclasp. Look the person in the eye and smile.

"How do you do?" or "I am very glad to meet you," are always appropriate acknowledgments.

Remembering names

Many people, on being introduced, fail to remember the person's name. This can be awkward enough in social situations, but can be damaging when dealing with the public.

To most people, the sweetest sound they hear is their own name.

Using a person's name while conducting business can be extremely helpful in creating a favorable attitude.

Be sure to get the name, write it down if necessary, and use it. If you don't hear the name, ask the person to repeat it, and **get it right!**

Shaking hands

When you are introduced to a person, you should shake their hand. (An exception is if your hand is quite dirty from your work.)

Correspondence

Although a letter is a substitute for a face-to-face contact, it should try to create just as favorable an impression as a personal contact. It should be friendly, brief and correct.

1. The letter should be addressed correctly. Spelling should be accurate, especially the name of the person being written to, and grammar always should be correct.

2. Get the attention of the person at the beginning of the letter.

3. Use their name.

4. Avoid the use of formal language and archaic expressions.

5. Refer to the other person's letter if there is one.

6. Use short simple words when you can. Many people make the mistake of writing in a different language than the one they speak. Ask yourself: "Does the word convey the exact shade of meaning I wish to convey?"

7. Do not waste a lot of time on a build-up to what you have to say. Just say what is to be said, answer the question to be answered, or ask what is to be asked. Then, close the letter.

8. Do not be discourteous in an attempt to avoid wordiness.

9. Be sure of the facts put in your letter. When you have the facts, label them as such. Beware of generalizations. Be as specific as possible.

10. Think of the reader as an individual. When you write your letter, keep in mind the person to whom you are writing. Try to anticipate questions that will come to their mind as a result of your letter, then answer those questions also.

Safety and public relations

Good public relations can be established in a way which sometimes is overlooked. The handling of motor vehicles is most important in field operations.

1. Both the best and, unfortunately, the worst impressions of the public frequently are made with public vehicles and the driving manners of volunteers. A person driving a city or fire district vehicle must lean over backwards to obey all traffic regulations. Extra courtesies paid to other drivers will pay off in good public relations.

2. By yielding the right of way, courtesy may become contagious. In the long run, this may make your city or district a safer place to drive for all citizens.

3. When volunteers park vehicles in the street, they have an added obligation. They must consider the flow of traffic and minimize as much as possible any interference with the traffic pattern.

4. Drivers of public vehicles play an important part in maintaining and improving the status of public relations. Vehicles have distinctive markings and are readily recognized. Volunteers who drive these vehicles and do not drive properly can cause great harm to the organization's public relations.

5. Every volunteer should make a special effort to be kind, courteous, and considerate when driving government vehicles.

6. Where a volunteer parks a vehicle is important. If a vehicle is parked in front of a tavern, grocery store, crosswalk, fire hydrant, no parking zone, or is double-parked, it will bring complaints from citizens.

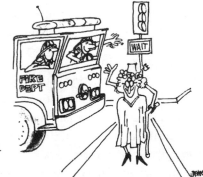

7. In driving a city or fire district vehicle, remember that courtesy equals safety. A courteous driver is also a safe driver. Drive Legally!

Rather than just doing a job for you or the department, a volunteer should understand that their work is being done because the citizens need certain services in return for their taxes.

If volunteers are encouraged to take pride in what they do, it will raise the morale of the work group.

What should be taught?

In addition to telephone courtesy, business etiquette, proper correspondence, and safety, the following items should also be included in your public relations training program:

1. **Organization of City or County Government.** Volunteers should be familiar with the history, organization, and operation of the city or fire district itself. They can then usually answer questions from relatives, friends, neighbors, and the many citizens they work with. When volunteers know about their city or district, they will take greater pride in their work and the organization they work for.

2. **Personality training.** Volunteers who come in contact with the public frequently must be especially careful in their jobs. Public opinion is strongly influenced by the behavior and attitudes of volunteers. When training volunteers, you should stress the importance of the attitude they display, the quality and quantity of information they furnish, the tone of voice used, and personal appearance. Teaching volunteers about answering questions and handling complaints are also desirable elements of training.

Developing programs

You can develop your training program in three separate steps as follows:

1. **Internal relations.** The first step should concentrate on relations between volunteers and their Officers or Supervisors. These would include the most common relations among volunteers in your company or department, relations with other personnel in the department, contacts with each other outside the department, supervisors inside and outside the department, and relationships between the volunteers and the city manager, mayor, commissioners, etc.

2. **External relations.** The next step involves direct and indirect relations which volunteers have with the public and importance of such contacts in the overall development of public opinions.

3. **Total relations.** The final phase of a training program could demonstrate how internal and external relations are inter-related. It is the combination of volunteer relations and contacts with citizens that result in overall public opinion regarding government.

Training your volunteers

Effective public relations do not just happen.

Volunteers do not automatically gain a keen sense of good public relations when they volunteer for the organization.

Training for better public relations means teaching volunteers to be more personally aware of their work.

Volunteers should feel they are serving the public.

Volunteers who come in contact with the public frequently must be especially careful in their jobs. Public opinion is strongly influenced by the behavior and attitudes of volunteers.

Five keys to good public relations

1. Good public relations are extremely important to supervisors.

2. Public relations include everything a city or fire district does which has any effect on public attitude.

3. Volunteer citizen contacts may include contacts in person, by telephone, or by correspondence.

4. Vehicle operation on public streets and roads is often overlooked as a place for citizen contact.

5. Officers need to train themselves and their volunteers in how to deal with the citizens in each of these areas.

The Art of Getting Along

Sooner or later, a person, if they are wise, discovers that business life is a mixture of good days and bad, victory and defeat, give and take. They learn:

- those who lose their temper usually lose.

- that all people have burnt toast for breakfast now and then and that they shouldn't take the other person's grouch too seriously.

- that carrying a chip on one's shoulder is the easiest way to get into a fight.

- that the quickest way to become unpopular is to carry tales and gossip about others.

- that buck-passing always turns out to be a boomerang and that it never pays.

- that everyone is human and that it doesn't do any harm to smile and say "Good Morning," even if it's raining.

- that most of the others are as ambitious as they are, that they have brains that are as good or better, and that hard work and not cleverness is the secret of success.

- to sympathize with the youngster coming into the business, because they remember how bewildered they were when they first started out.

- not to worry when they lose an order because experience has shown that if they always gives their best, their average will break pretty well.

- that bosses are not monsters trying to get the last ounce of work out of them for the least amount of pay, but that they are usually fine people who have succeeded through hard work and who want to do right things.

- that the gang is not any harder to get along with in one place than another, and that "getting along" depends about 98 percent on their own behavior.

Organizations and departments are made or broken by their Public Relations – much is said about Public Relations but very little is printed that really defines what we mean by Public Relations.

The term Public Relations is used to explain the relationship of the public to an organization or business. Our Public Relations are either good, fair, or bad, depending upon the opinion the public has of us, at any given time.

Public Opinion is formed by observing or hearing about acts or actions of the organization or one or more of its members.

Public Opinion is formed by observing or hearing about acts or actions of the organization or one or more of its members.

The mental disposition of the observer is a factor in forming Public Opinion. What will bring favorable opinion at one time will bring unfavorable opinion at another time under different circumstances.

We must constantly be aware of our public exposure. Our acts and actions are transferred into Public Opinion, like it or not.

Public Opinion can be controlled, but it requires planning and constant effort. Each act must be evaluated, each statement must be weighed. We must be alert to any opportunity to improve our Public Relations. Each member of the organization is its Ambassador to the public – **you are** the organization or the business when you represent it to the public. Public Opinion is formed about **you** and transferred to the organization.

Keep your best foot forward and be sure the shoe is shined.

Notes ■

Customer Service And Marketing ■■■■■■ 18

Sharp Contrast Between Fire Service and Private Sector

I'm always amazed at the sharp contrast that exists between the fire service and the private sector. Imagine for a minute a private company like Federal Express driving down the road with the side of their truck completely bare, except for a 4-inch circular logo on their door that requires a map and a set of instructions on how to read the logo in order to figure out who the truck even belongs to. As crazy as that sounds, that's what fire departments do every day. A fire engine on the road is probably the single greatest attraction bystanders and other drivers will notice during their morning commute. Firefighters and fire apparatus draw a tremendous amount of attention. Yet, the compartment doors on the side of fire engines often stand empty.

The fire service should take a lesson from the private sector on how to advertise who we are, what our product is and, more importantly, develop some loyalty between our customers and our fire department. If you are one of those firefighters that thinks you work for "the fire department" then we have some work to do. Each of us should identify with our fire department, such as Tualatin Valley Fire & Rescue or Sutherlin Fire Department because absent that ownership, our customers will not care who provides their service, just so "a" fire department does. Most firefighters have a tremendous amount of pride in their fire department and the people they work around and their ability to do their job. However, we do such a poor job of marketing that our customers become accustomed to thinking of us as "the fire department." The absence of product loyalty will destroy any business. Imagine someone going to the store to buy a pair of tennis shoes and not really caring whether they are Nikes or Reeboks or going to the auto mart to buy a car but not really caring what brand they purchased. Businesses would go broke and there would be huge opportunities for their competition to create loyalty and a preference for a product brand or type in the minds of the consumer.

The public sector has left their customers without product loyalty and without creating a relationship between them and their customers simply because of complacency. Marketing your fire department can help establish that relationship and loyalty between your customer and your fire department. One of the fundamental elements that must occur first, however, is accepting the fact that the people we serve are in fact customers. It's a confusing point. They don't look like customers, because they don't hand us money when we serve them. They don't come to our place of business and exchange money for a service on the spot. But rather, the exchange of money is out of sight of the people that provide the service. When people pay their taxes, they are in fact buying a service. They are buying our service. They are buying protection from fires, public assistance and rescue services. The fact that no money changes hands is a moot point.

The fact is, people pay for our service, we provide services to them and since competition does exist in the fire service, we are literally competing for customers. Many states in this country are witnessing battles between the public and the private

> *The fire service should take a lesson from the private sector on how to advertise who we are, what our product is and, more importantly, develop some loyalty between our customers and our fire department.*

sector. And what's even more evident are the battles between two public sector fire departments. In every county in this country, an annexation battle has been playing out for years. One department is gaining ground while the other is losing ground. The departments that are losing ground typically understand product loyalty, name identification and marketing because their only security lies in the fact that their customers prefer their service over the service of someone else.

Tualatin Valley Fire and Rescue is home to some of the world's largest corporate giants including the world headquarters of NIKE. We could learn some important lessons about marketing from NIKE. How to do it and how not to do it. NIKE knows that there is a direct correlation between sales and product loyalty, and product loyalty starts with name and logo identification. Now for the fire service, logo identification is not difficult. We're the people that ride around in those big fire trucks. On the other hand, that makes you "just a firefighter" or "just the fire department," when in fact you want to be known as "your fire department." Name inserted. NIKE puts their logo or name on virtually every piece of clothing, duffel bag, luggage or any product they have. Even the trucks hauling their products are clearly marked with the NIKE swoosh. Kodak was once the most recognized symbol in the world. Even absent the name Kodak, their logo was more identifiable than any other logo.

The fire department misses many opportunities to market. Let's start with the side of a fire apparatus. The name of your fire department should be in 12" to 18" block letters in 3-M scotch light across the full length of your compartment doors. You should also put the most important safety message you have to communicate on the sides of your fire apparatus in reflective tape. "Working smoke detectors save lives" is the most important thing we can put on the side of our fire apparatus and staff cars. Because our fire apparatus attract so much attention, take the opportunity to tell people who you are and what the most important thing you have to tell them is. The other opportunity to market ourselves is in front of the fire station. Because most people drive by and look at the fire station, we should put a reader board in front of our station. This reader board should be used to recognize volunteer's contributions, communicate your weekly safety message to passersby or potentially recruit new members. The uses are endless.

What we wear is also an excellent opportunity to market ourselves. Our turnout coats should bear our fire department's name, as should an easy to read shoulder patch, the back of our T-shirts in large, block white lettering, our sweatshirts, jackets, coats, trauma boxes and fire helmets. There would be virtually no angle a photographer could take our picture while operating an emergency incident where our name escapes their view. Many public relations experts have known for years the value of 15 to 30 seconds on the evening news or the name clearly identified on the front cover of the evening paper. Constant display of your fire department's name provides for "Top of the mind awareness." Top of the mind awareness is valuable when people are deciding whether to support your tax request or your request for a new piece of fire apparatus and it has an impact on how they perceive your organization.

Professionally displayed names, logos, properly maintained equipment and properly dressed staff all project a very professional image and provide a visual reassurance to people that they are in fact getting their money's worth and purchasing the highest quality services possible. Those departments that are exclusively from contributions by degree of visibility and professionalism build "esprit de corps" and pride in the organization. A key element to a good marketing program is the public information officer. Each organization should have someone assigned to the role of Public Information Officer (PIO).

The PIO's role is to communicate information about emergencies, departmental activities, and policies to the media and to the community, while at the same time, their key responsibility is to seek out and capture public relations opportunities for the department. After every emergency incident of significance, the PIO should craft a press release, send it to the local media and follow that up with a phone call. By being the first to communicate the story to the media, you can more effectively control their perception of the event, positive or negative, rather than hearing about a potentially negative event and forming opinions which you then have to completely counteract and overcome. It's almost impossible to overcome negative impressions once they have their mind made up about what "the truth is." The public information officer should also coordinate public speaking events for the chief officer in the department.

A local speakers bureau which outlines the topics and the individuals who are available to talk to the local Lions and Rotary Club or any other group who has an interest will create tremendous opportunities for you to contact your community. Although these daily contacts seem insignificant, they often lead to public/private partnerships and local civic organizations asking you how they can help you serve the community. In Sutherlin, OR, Fire Chief Tom Wells created a partnership with the local Lions Club and high school which lead to a complete hydraulic rescue tool and accessories being donated to the fire department and lead to an annual smoke detector inspection day where the fire department along with the Lions Club, seniors group and high school graduating class literally goes door to door across the entire town of 6,000 people checking and installing free smoke detectors and batteries in the homes of everyone that needs them. The local golf course provides golf carts for transportation and the local insurance company provides the detectors and batteries free of charge. This is an example of a marketing opportunity and, more importantly, an effective fire prevention and public education tool.

During the weekend of the smoke detector inspection check, the entire community is reminded of the need for smoke detectors and the importance of the detectors by virtue of the entire fire department and civic community turning out to eliminate this single fire hazard in people's homes. The participant's are known as Detector Inspectors and are given T-shirts that identify them as such. After the entire town has been completed, the local Abby's Pizza Inn treats everyone to pizza and pop which has historically been donated. These kinds of partnerships form lasting relationships and commitments which in and of themselves have public relations value. Imagine your city council saying no to the fire department's request for improved equipment or stations, when virtually every business and civic leader in town thinks the fire department is the most impressive organization since the formation of the American Red Cross.

Public Information Officers must also be attuned to the different types of media and their functions. There are essentially four types of media: television, print, radio and electronic.

Let's imagine for a moment that you have a disaster of significant magnitude to draw in all four types of media into your community. A well-trained public information officer knows that media needs access to the scene. Access does not mean that they can wander aimlessly about the emergency scene, but it does mean that if you fail to provide them an adequate vantage point to see the incident from a safe distance, they will find a way around your security line and place themselves in danger in order to get the story.

The PIO should work with the incident commander to locate the safest point close to the incident where media can be taken to view the incident. This point should be closer than the general public is allowed, but not as close as firefighters are allowed. The television crews in this case, specifically the cameramen for television crews, want action shots. TV is a very visual medium. They are looking for action, color and a unique perspective. Given four or five TV stations, not all of them will want the identical shot on the evening news and care should be taken to provide them each something unique for their storyline that evening. Newspaper photographers have the same interest as television cameramen. They all want an action shot that in and of itself explains what is going on at the incident. The reporters for both television and print want to capture some detail. The television reporter is going to want to talk briefly to you before placing you on camera to discuss the emergency incident. What you discuss off the camera is usually what's talked about during your interview. The camera person will often stand off to one side of the camera and ask you questions, and you should keep your answers short and to the point. Do not look at the camera, look at the interviewer, and talk to that person as if you're having a casual conversation. By keeping your sentences short, the editor is able to make your interview as short or as long as is necessary. Remember that they may use your voice only to dub over pictures that the cameraman has shot from their vantage point.

Television is thought to be a headline medium. Not much detail is provided through television. Their purpose is to pique your interest and cause you to be interested in the story. Their priorities are to break a unique story and to do it first. Print medium on the other hand, newspapers, are interested in detail. Given a newspaper reporter and a television reporter arriving on the same emergency scene simultaneously, care should be taken to address the TV videographer and print photographer's needs first, followed by an interview with the TV reporter and then taking time to meet with the print reporter. The print reporter is going to need a tremendous amount of detail. That's the strength of newspapers.

Take some time to get as much information as you can before talking to a print reporter, as they will need more information than any other medium. Radio reporters are not interested in a picture of the scene. They are interested in the sounds of the incident as well as a taped interview with you about what's happening as the scene. Radio can be as fast or faster than TV and can relay more detail in a short period of time because radio requires so little editing. A feature of radio is that it can be done over the telephone and you are often asked to do your interviews by telephone and placed directly on the air. As with TV, be sure to speak in short sentences and provide the editor and the interviewer lots of opportunities to edit your interview to their time format.

The web or other electronic news mediums are now becoming more prevalent. The internet is becoming a tremendous tool for the PIO. Press releases can be posted and updated from any location at anytime and can be instantly accessed by multiple news agencies and citizens.

Media is also an effective tool in creating an image about your fire department. Not all fire departments have the same image. Some are viewed as efficient, some are viewed as heroic, others yet are viewed as lazy and inefficient. It all depends on the way that the media has characterized your organization to the community and what kind of a relationship you have with the community.

Image is important; in fact many businesses would tell you image is everything. If we re-examine the NIKE model, we would recognize that they never talk about whether their shoe is superior to Reebok or not. They market their product by having the world's greatest athletes wear NIKE. Athletes who want to be associated with the success of the featured NIKE athlete will draw the conclusion that the NIKE shoes or products help the athlete perform and will also want to have that "performance edge" the professional athlete does.

Firefighting is the most respected occupation in the United States, according to a 1994 poll. What's worrisome is that no business, athletic team or individual that's ever achieved No. 1 status has stayed at the No. 1 position. Our ability to stay the most respected occupation in the country will depend on two things: how well we do our job and how well we market ourselves and create an image in the minds of the taxpayers.

Let's examine some things that impact our image. Rude and overly aggressive driving practices can hurt our image. Driving manners when driving emergency response apparatus can be frustrating. Quiet vehicles and loud music often create circumstances where the drivers of other vehicles cannot hear us coming and the increased pressure from other traffic causes them to not pay attention as other vehicles move around and up behind them. All these things can lead to frustration. The professional engineer and fire officer does not give in to the temptation of driving in a rude manner or displaying their frustration to other drivers.

How we comport ourselves as firefighters is also critical. A comportment is not the thing inside of a fire apparatus: that's a compartment. A comportment is, "Do you look like the job you're there to do?" Looking professional is a big part of having people perceive you as professional, which is a fundamental requirement of building your image. Many firefighters claim that it doesn't matter how they look, just so they do their job well. Although job performance is critical to maintaining a positive image, it's short sighted to think that you personally serve enough people to create an image of your organization's professionalism.

For every person you serve literally hundreds or thousands of people will see you driving to the grocery store, on the drill ground, on and off duty and they will form an opinion about your professionalism based on how you look and how you conduct yourself. The airlines figured this out many years ago when they required their pilots

Media is also an effective tool in creating an image about your fire department.

to wear very uncomfortable but professional uniforms as they fly their planes. An argument could be made that you are a better pilot if you're comfortable when you're operating the plane. And if you are comfortable your concentration will be better and your performance will increase. So why is it that airlines require their pilots to dress so professionally? And further, why do they require them to walk past the passengers rather than enter the airplane through the same service entrance that the maintenance and food services employees do? It's simple. They do it for image. A professionally dressed and comported pilot.

The key element to having a positive professional image is conducting ourselves in a professional manner, although the unique and often funny circumstances that present themselves in emergency scenes can be good cause for jokes and laughter between emergency responders. The fact remains that it's rarely appropriate and we seldom want our customers to hear what we say during these private moments between emergency responders. Bystanders and media are always drawing conclusions about our professionalism based on how we act during and after the emergency scene. Very few firefighters and emergency responders can say that they haven't made this mistake.

Many believe the most neglected, yet most important, rule of professionalism and building a positive image is the attitude of our personnel. Everyone deals with stress differently and the emergency scene can create stresses and anxieties beyond what the normal person deals with in a lifetime. We must keep our attitudes and opinions to ourselves during these emergency operations and make sure that we are always treating people with compassion and the way we would like our families treated in the same situation. A poor attitude or rude behavior is never acceptable. If you witness this occurring, you should remove your co-worker or subordinate from the immediate scene so that the customer does not suffer the ill effects of our frustration and/or stress.

Positive image is not just about what you should do; it's also about what you shouldn't do.

Here are a few "nevers" of public relations:

- Never intentionally mislead the public.

- Never speak negatively of your organization.

If you speak negatively of your organization, then people question your judgment for continuing to work there if it's such a bad place. They even begin to suspect that maybe your confidence won't allow you to find a job with a reputable fire department. Speaking negatively of your organization only causes people to question your judgment and your professionalism.

- Never air your dirty laundry in front of the public for political gain.

People who air their dirty laundry in the media or in front of the public tend to end up wearing someone else's dirty clothes. Once you make public "Dirty Laundry" it takes on a personality of its own. It soon achieves status well beyond your control and, more often than not, these issues will turn around and work to your disadvantage rather than your advantage.

- Never attempt to cover up if your organization has made a mistake. Admit it!

> *The key element to having a positive professional image is conducting ourselves in a professional manner.*

> *Many believe the most neglected, yet most important, rule of professionalism and building a positive image is the attitude of our personnel.*

> *Positive image is not just about what you should do; it's also about what you shouldn't do.*

Legal advice may direct otherwise, however, there are two types of victories. One occurs in the courtroom and one occurs in the eye of public opinion. You may legally have every right to remain silent on the issue if your fire department has made a mistake, and technically, there may be legal reasons why you should remain silent. Conversely, if the public believes that you are covering up, or if the general public can see that you've made a mistake (and it doesn't take an expert to draw that conclusion), however you refuse to explain your actions, you can lose the credibility of the fire department forever. No courtroom victory will ever restore that. Remember that your legal counsel is retained to advise you on maximizing your legal position and your legal advantage. It is not necessarily the same as the right policy decision.

• Never burn a bridge with the media.

A good rule of thumb is: don't mess with someone who buys ink by the barrel and paper by the ton. Worse yet, video tape and audio tape are reusable. We must ask ourselves if the Rodney King incident would have been the issue it was today had the issue not been video taped.

Can you think of a situation where it would have been tremendously detrimental to your organization, had it been videotaped? Once an organization establishes a negative and distasteful relationship with the local media agency, they can "make you their hobby." Being the hobby of a reporter or a media agency that sees your organization as wasteful, incompetent or just plain pathetic will never result in a positive outcome as long as the fire department and the local newspaper are busy throwing darts at each other. Building positive media relationships starts with telling the truth, being open and honest and having ongoing relationships even when there aren't any issues to deal with that day.

Complaints

We have a simple philosophy around complaints. In the mid-70's, a friend worked for Tandy Corporation, as a store manager. His immediate supervisor was the District Manager who happened to have his office in the same community. One day, the district manager called him into his office to explain that he failed the company by making a customer so angry that they vowed they would never return. As he pled his case to the boss, he shook his head in disappointment. He said he didn't get it. Turns out, he was right. He had lost a customer forever because he had followed the company policies. But, the company culture was such that he could throw out policies to save a customer. To the company, the customer is never worth losing.

Managers all had latitude to correct situations that were outside policy. His district manager made it all too clear. "Be a hero at your own level because if it gets to me, I'll be the hero and then I'll send your store the bill," he said. "Which comes out of your bonus, in case you didn't know." In the fire service, we often fail to empower our officers to be heroes in matters other than managing the emergency. That is something we should all work to correct.

Have you ever asked yourself why we grant firefighters and paramedics the authority to literally place people's lives in their hands, inject life altering drugs into their veins, but if they were to turn around with the attic ladder as they carried it through the house and break a front window out, we no longer trust them to decide

Don't mess with someone who buys ink by the barrel and paper by the ton.

Be a hero at your own level.

that they made a mistake, pick up the telephone and call a 24-hour glass repair place to come and fix the damages they created? Why is it that we have this divergence of trust as it relates to our employees? Being a hero at your own level means pushing authority and responsibility as low in the organization as you possibly can, enabling your personnel at the lowest level to solve customer's problems. After all, we are in the help people business. For those of us that still believe we're in the fire business, we want you to go back and look at last year's statistics and see exactly how many working structure fires you fought last year.

The reality is, we're a help people service that also fights fire. We just happen to be the only people that fight fire, which makes it our market niche, but nonetheless, we need to train our employees and volunteers on the techniques used to field, handle and resolve complaints at the lowest level possible. And that means empowering our employees to solve problems.

There are three ways to handle a complaint:

- First, is the immediate resolution. That's when the first person that receives the report resolves the complaint.

- The second way to handle a complaint is an intermediate resolution. That's where a complaint is handled by the second, or at the most, the third party whom the customer comes in contact with. In this process the complainant is identified and a resolution plan and timetable are discussed and the original complaint receiver passes the customer to the second or third party, only after briefing the second or third party. So when the contact is made, the customer's name and problem are known by the party resolving the complaint.

- The third way to handle a complaint is to have a permanent enemy. Remember, friends come and go, but enemies accumulate. You can accumulate enemies by passing a dissatisfied customer from person to person in your organization or by inconveniencing your customer by making them come down to your office and fill out a bunch of paperwork or a host of other aggravating, buckpassing techniques. Have you ever tried to return something at a local store only to be passed from person to person or stand in line for an hour? By the time you finally get your problem resolved you still hate the place and you vow never to return. Had they solved your problem early, treated you like you were a customer standing there with a thousand dollars in your hands, your attitude would have been completely different.

You must deal with complaints in a rapid and expeditious manner. Some tips for resolving complaints:

- Clearly understand the complaint. Do perception checking.

- Understand the motive of the complainant. People aren't always interested in money. Sometimes they are just interested in you listening to their grievance.

- Do not act defensive or territorial. Give the clear impression that you will do everything you can to resolve the problem. Assure the complainant that you will personally handle the problem, even if it that means passing it to another person.

Recruiting, Training, and Maintaining

Make sure you deliver the complainant to the person that can solve the problem.

- Be sure to ask, "What will it take to make you happy?"

- Act swiftly in returning calls, hearing and acting on any complaint.

- Resolve the issue as quickly as possible after receiving it, and remember, be a hero at your own level.

Public relations is not some mystical science. It's about treating your customers the way you would like your family treated when you're out of town. It's about the golden rule, treating others as we would like to be treated ourselves. It's going the extra mile, being a hero at your own level and creating a relationship between your fire department and the customers, so that they want your fire department, not just a fire department. We can never forget that firefighters are the basis to all communication with the public, positive or negative. Firefighters are heroes to the public. Our attitudes and actions can build or destroy that relationship and perception. Common courtesy, professional service and a healthy attitude are the three best public relations tools available to any department.

Public relations is not some mystical science. It's about treating your customers the way you would like your family treated when you're out of town.

References

Snook, Jack & Olsen, Dan. *Recruiting, Training and Maintaining Volunteer Firefighters*, M.D.I., Inc., 1982.

Snook, Jack. *Volunteerism in America - Past, Present, and Future*, M.D.I., Inc., 1982.

Perkins, K., Ph.d., *Volunteer Firefighters in the United States: A Sociological Profile of America's Bravest*, 1987.

Rauner, Judy, *Helping People Volunteer*, Marlborough Publications, 1980.

Wilson, Marlene, *The Effective Management of Volunteer Programs*, Vol. Man. Assoc., 1976.

Jack W. Snook

Jack W. Snook is the President of Emergency Services Consulting inc. (ESCi). They are an international firm providing specialized and high quality fire, police, communications, and emergency medical consulting services to organizations throughout the United States and Canada. Mr. Snook has been with the company since 1976. He has over 29 years of private and public sector experience at multiple levels. His formal education includes a Masters degree in Public Administration, a Bachelor of Science degree in Fire Administration, and an Associates of Arts degree in Fire Science. His career ranges from being the chief executive officer of a city and corporation to being the chief and administrator of one of the nation's eighty largest fire departments. He has been looked upon as a national leader in the fire service for many years.

Mr. Snook has extensive experience in providing consulting services to clients throughout the world. Areas of expertise include management and organization reviews, cooperative service and consolidation, strategic planning, program evaluations, and risk assessment. In 1991, he signed an exclusive contract with the International Association of Fire Chiefs (IAFC) to present all of their cooperative effort workshops nationwide. He is the co-author of the book entitled *Making the Pieces Fit, Through Cooperative Effort*. The publication is the recommended reference book by the IAFC and the International City/County Managers Association (ICMA). He has served as the project manager for over a dozen of ESCi's cooperative effort studies throughout North America.

Recent assignments and/or appointments include facilitator of the National Fallen Firefighters Foundation national summit to reduce line of duty deaths in America; moderator of the nation's first symposium to bring healthcare officials and the fire service together to develop a model program to reduce healthcare facility deaths and injuries; facilitator of a national roundtable to discuss emerging codes; and facilitator of the International Association of Fire Chief's strategic plan (two years). Mr. Snook has key-noted over seventy-five conferences and conventions and has spoken at over 200 events.

Jeffrey Johnson

Jeffrey Johnson is the Fire Chief of Tualatin Valley Fire and Rescue. He joined the Fire District in 1989 from a fire service career in Douglas County, Oregon. Chief Johnson served as Division Chief and Assistant Chief at TVF&R before becoming Fire Chief in1995.

During his administration, TVF&R has become an accredited agency and added three additional cities and two additional fire districts to its service area, bringing the District's resident population to over 400,000. Spanning three counties, the District is a combination department that includes over 400 career firefighters and support staff, and approximately 100 volunteers. TVF&R was recently honored with the International Association of Fire Chiefs' (IAFC) top Fire Service Award for Excellence.

Chief Johnson has completed the Executive Fire Officer Program. He holds a Bachelor of Science degree (Magna Cum Laude) in Business Management and Communications, as well as Associate degrees in Fire Science and Criminal Justice Administration. He is past President of the Board of Directors of the Oregon Fire Chiefs Association (OFCA), on the Board of Directors of the Western Fire Chiefs Association, Chair of the Governor's Fire Policy Council, and a member of IFSTA's Development & Validation Committee. He is also active in various local chambers and business associations.

Dan C. Olsen

Dan Olsen's fire service career includes serving in positions from the rank of Firefighter to Fire Chief. He has worked with both career and volunteer personnel since joining the fire-rescue service in 1972. His administrative assignments have included Chief Fire Marshal for Clark County (Washington), Fire Chief for the City of Milwaukee (Oregon), and Fire Chief for Marion County (Oregon) Fire District No. 1. Currently, he is the Fire Chief for South Lane County (Oregon) Fire & Rescue.

He and Trish, his wife of 34 years, live in Oregon and have one son, Thad, who is a career fire officer.

John M. Buckman, III

John M. Buckman, III is the Special Projects Manager in the Division of Training for the Indiana Department of Homeland Security.

Chief Buckman served as fire chief of the German Township Volunteer Fire Department in Evansville, Indiana, and is past president of the International Association of Fire Chiefs. In 2001, he received the designation Chief Officer from the Commission on Fire Accreditation International. Chief Buckman graduated from the Executive Fire Officer Program at the National Fire Academy in 1988. President George W. Bush appointed Chief Buckman to the Department of Homeland Security State, Tribal, and Local Advisory Group. In 2000 he was appointed to the America Burning Revisited Commission by President William J. Clinton. In 1995, *Fire Chief* magazine named Chief Buckman Fire Chief of the Year. He has authored over 70 articles and presented at a variety of conferences in all 50 states, as well as Canada.

Notes